AF870797

Trends in Terahertz Technology

Edited by Yahya M. Meziani and Jesús E. Velázquez-Pérez

Published in London, United Kingdom

Trends in Terahertz Technology
http://dx.doi.org/10.5772/intechopen.102214
Edited by Yahya M. Meziani and Jesús E. Velázquez-Pérez

Contributors
Joshua O. Island, Randy M. Sterbentz, Taiichi Otsuji, Robin-Cristian Bucur-Portase, Seiji Kojima, Salman Alfihed, Abdullah Alharbi, Mohammed Alhamdan, Naif Alshamrani, Hadba Hussain, Osamu Wada, Doddoji Ramachari, Chan-Shan Yang, Takashi Uchino, Ci-Ling Pan

© The Editor(s) and the Author(s) 2023

The rights of the editor(s) and the author(s) have been asserted in accordance with the Copyright, Designs and Patents Act 1988. All rights to the book as a whole are reserved by INTECHOPEN LIMITED. The book as a whole (compilation) cannot be reproduced, distributed or used for commercial or non-commercial purposes without INTECHOPEN LIMITED's written permission. Enquiries concerning the use of the book should be directed to INTECHOPEN LIMITED rights and permissions department (permissions@intechopen.com).

Violations are liable to prosecution under the governing Copyright Law.

(cc) BY

Individual chapters of this publication are distributed under the terms of the Creative Commons Attribution 3.0 Unported License which permits commercial use, distribution and reproduction of the individual chapters, provided the original author(s) and source publication are appropriately acknowledged. If so indicated, certain images may not be included under the Creative Commons license. In such cases users will need to obtain permission from the license holder to reproduce the material. More details and guidelines concerning content reuse and adaptation can be found at http://www.intechopen.com/copyright-policy.html.

Notice

Statements and opinions expressed in the chapters are these of the individual contributors and not necessarily those of the editors or publisher. No responsibility is accepted for the accuracy of information contained in the published chapters. The publisher assumes no responsibility for any damage or injury to persons or property arising out of the use of any materials, instructions, methods or ideas contained in the book.

First published in London, United Kingdom, 2023 by IntechOpen
IntechOpen is the global imprint of INTECHOPEN LIMITED, registered in England and Wales, registration number: 11086078, 5 Princes Gate Court, London, SW7 2QJ, United Kingdom

British Library Cataloguing-in-Publication Data
A catalogue record for this book is available from the British Library

Additional hard and PDF copies can be obtained from orders@intechopen.com

Trends in Terahertz Technology
Edited by Yahya M. Meziani and Jesús E. Velázquez-Pérez
p. cm.
Print ISBN 978-1-83768-167-9
Online ISBN 978-1-83768-168-6
eBook (PDF) ISBN 978-1-83768-169-3

We are IntechOpen, the world's leading publisher of Open Access books

Built by scientists, for scientists

6,600+
Open access books available

177,000+
International authors and editors

195M+
Downloads

156
Countries delivered to

Our authors are among the
Top 1%
most cited scientists

12.2%
Contributors from top 500 universities

WEB OF SCIENCE™

Selection of our books indexed in the Book Citation Index
in Web of Science™ Core Collection (BKCI)

Interested in publishing with us?
Contact book.department@intechopen.com

Numbers displayed above are based on latest data collected.
For more information visit www.intechopen.com

Meet the editors

Yahya Moubarak Meziani received his Ph.D. in 2001 from Montpellier 2 University, France. He worked there as a research assistant and teaching assistant from 2001 to 2005. He was a Japan Society for the Promotion of Science (JSPS) post-doc fellow in 2005–2006 and an assistant professor for the Research Institute of Electrical Communication, Tohoku University, Japan, in 2006–2008. In 2008, he was awarded the Spanish Ramón y Cajal tenure track at Salamanca University, Spain. Since 2013, he has been an associate professor at Salamanca University. He has authored more than 70 papers and 3 book chapters and participated in more than 100 international conferences. He is co-editor of the journal Electronics and was granted 30 competitive projects as IP and a member of the investigation team. His interests include terahertz technology, imaging and inspection, sensors, field effect transistors, and plasmonic rectification.

Jesús Enrique Velázquez-Pérez obtained his Ph.D. from the Universidad de Salamanca (USAL), Spain, and the Institute of Fundamental Electronics, University of ParisXI, Orsay, France, in 1990 with his work on the simulation of high-speed devices using models based on the Monte Carlo (MC) technique. In 1990, he joined the faculty of USAL as a Professor of Electronics. At that time, his research focused on Radiofrequency and noise modeling of heterojunction bipolar transistors (HBTs). From 1998 to 2004, he was an invited researcher at the Imperial College London, UK, developing n-channel Si/SiGe MODFETs technology for low-power RF and millimeter-wave applications. This work led to strained-Si MODFETs and, later, to graphene field-effect transistor (FET) plasma-wave-based THz detectors. He has coauthored about 100 papers and more than 150 communications to international conferences.

Contents

Preface

Most of the electromagnetic (EM) spectrum has been extensively studied and used in countless industrial applications in communications, microscopy, spectroscopy and material analysis, medicine, and so on. Nevertheless, until a short time ago, the portion of the EM spectrum ranging from 100 GHz (3 mm) to 10 THz (30 μm), commonly known as the Terahertz (THz) region, has remained elusive for experimental science. The THz region of the electromagnetic spectrum is the least explored, the least well-known, and the last to be put to commercial use. The first measurements in the THz region were carried out by Heinrich Rubens in 1900 who measured the emission spectrum of the blackbody down to a few THz. His study played a key role in the formulation of the Max Plank blackbody radiation law.

Nevertheless, over the ensuing seven decades, the THz region received little attention. It was considered an "exotic" research field despite its recognized wide range of potential applications. It was only in the 1970s, once it was clearly demonstrated that most of the information about interstellar dust precisely lies in the THz portion of the EM spectrum, that research in THz was tackled vigorously. Since then, many telescopes have been designed and built to operate in the THz region around the world, especially in space-borne observatories. This has made available THz instrumentation that subsequently has enabled the progress of THz science in many research fields beyond astronomy. The interest in THz science also increased due to some unique properties of THz radiation, including imaging of concealed objects, spectroscopy (various rotational, vibrational, and translational modes of lightweight molecules are within the THz range, i.e., many substances have a THz fingerprint), communications with a bandwidth significantly larger than those based on microwaves, and more.

The past two decades have seen tremendous advances in the availability of THz sources and detectors that paved the way for commercial use, which in turn have fueled research in new applications of THz radiation. Some examples of new THz sources are Quantum Cascade Lasers (QCLs), photoconductors and photodiodes, nonlinear optical (NLO) materials excited by lasers, electronic sources, and oscillators. Similarly, new detectors based on solid-state devices such as Schottky-barrier diodes and bolometers, plasma waves in field-effect transistors, and others have been demonstrated.

This book includes six chapters summarizing the most recent progress and efforts in different research areas of THz science. It is organized into three sections: "Terahertz Time-Domain Spectroscopy," "2D Materials for Terahertz Technology," and "Biology and Terahertz Radiation."

The first section, "Terahertz Time-Domain Spectroscopy", consists of three chapters. Chapter 1 "On-Chip Sub-Diffraction THz Spectroscopy of Materials and Liquids" by Randy M. Sterbentz and Joshua O. Island, focuses on the use of fast photoconductive switches to generate and detect on-chip THz pulses using a femtosecond laser to study the picosecond dynamics and THz spectroscopy of materials and liquids. Chapter 2, entitled “Terahertz Spectroscopy in Advanced Materials Science” by Seiji Kojima, reports on the recent progress of vibrational spectroscopy using THz time-domain spectroscopy, THz time-domain spectroscopic ellipsometry, and far-infrared spectroscopy in advanced materials science on glassy and crystalline pharmaceuticals, ferroelectrics, and polar metallic materials. Chapter 3, “Systematic Characterization of High-Dielectric Constant Glass Materials Using THz-TDS Technique” by Osamu Wada, Doddoji Ramachari, Chan-Shan Yang, Takashi Uchino and Ci-Ling Pan, presents a study of oxyfluorosilicate (OFS) glasses by THz time-domain spectroscopy. High values of the dielectric constant (8–13) and low losses (6–9/cm) were found; a unified single oscillator-based model is used to explain the physical origin of the dielectric constant enhancement.

The second section, “2D Materials for Terahertz Technology”, includes two chapters. Chapter 4, “Two-Dimensional Materials for Terahertz Emission” by Abdullah Alharbi, Naif Alshamrani, Hadba Hussain, Mohammed Alhamdan and Salman Alfihed, focuses on 2D material-based emitters and discusses the physical principle of photoconductive THz emitters, the materials’ properties, and the research trends in the fabrication and characterization of THz sources based on 2D materials. Chapter 5, “Graphene-Based Plasmonic Terahertz Laser Transistors” by Taiichi Otsuji, presents recent advances in the research of graphene-based plasmonic THz laser transistors. Single-mode THz emission as well as broadband THz amplified spontaneous emission from graphene-channel field-effect transistor (GFET) laser chips is theoretically and experimentally demonstrated. To improve the laser performance, the use of graphene Dirac plasmons (GDPs) is proposed and experimental results clearly demonstrate room-temperature amplification of stimulated emission of THz’s radiation.

The third section, “Biology and Terahertz Radiation” includes the final chapter. Chapter 6, “Introduction to the Biological Effects of Terahertz Radiation” by Robin-Cristian Bucur-Portase, examines how THz radiation affects biological organisms via thermal and non-thermal effects ranging from modifications brought to protein activity to epigenetic changes that lead to altered metabolism or reproduction. The chapter discusses THz ability to enhance actin polymerization and alter gene expression.

Yahya M. Meziani and Jesús E. Velázquez-Pérez
Department of Applied Physics,
University of Salamanca,
Salamanca, Spain

Section 1

Terahertz Time-Domain Spectroscopy

Chapter 1

On-Chip Sub-Diffraction THz Spectroscopy of Materials and Liquids

Randy M. Sterbentz and Joshua O. Island

Abstract

This chapter summarizes the trends in terahertz measurements on the surface of rigid and flexible substrates. It focuses on research incorporating fast photoconductive switches to generate and detect on-chip THz pulses using a femtosecond laser. The chapter aims to review progress toward the study of picosecond dynamics and THz spectroscopy of materials and liquids. We emphasize general sub-diffraction techniques for THz spectroscopy, transmission line and waveguide design considerations, time-domain measurements for studies of material dynamics, and provide a survey of recent research on the THz spectroscopy of materials and liquids on-chip. We conclude with an outlook on the field and highlight promising new directions. This chapter is meant to be an introduction and a general guide to this emerging field for new researchers interested in on-chip THz studies.

Keywords: on-chip, terahertz, spectroscopy, materials, liquids

1. Introduction

Terahertz (THz) time-domain spectroscopy has led to a deeper understanding of the properties of matter. Developed from early measurements of the optical switching of silicon [1, 2], generation and detection of free-space THz pulses in the late 1980s [3–5] gave way to a new spectroscopic approach for fundamental science [6]. With both amplitude and phase sensitivity, the technique allows direct extraction of the complex-valued index of refraction. The THz spectrum covers a window of energies (and timescales) that host rich phenomena in solids including: important carrier dynamics [7, 8], charge ordering [9], gapped excitations in superconductors [8, 10], magnetic excitations and dynamics [11], and energy quantization in the quantum Hall effect [12–14].

THz radiation has a number of appealing properties for biomedical sciences as well [15–19]. The non-ionizing and non-destructive nature of the radiation makes it particularly promising for non-invasive, *in vivo* imaging and cancer detection [20–22].

The time resolution of time-domain measurements opens the door to studying important hydration dynamics in solvated biomolecules and chemical reactions at the picosecond timescale [23]. Moreover, the capacity of this frequency of radiation to resonantly excite intra- and inter-molecular oscillations presents a unique advantage in the detection of low-concentration biomolecules. Highlighting a few notable recent examples, Cheon et al. showed that methylated DNA presents a resonant excitation at 1.6 THz and the degree of methylation can be quantified by the amplitude of the absorption peak [24]. Niessen et al. have reported clear resonant features of lysozyme, E. coli DHFR, and RNA G-quadruplex [25].

While the majority of research involving THz spectroscopy has been done using free-space radiation, there is a growing interest in probing samples smaller than the diffraction limit on the surface of a chip. The diffraction limit puts a fundamental restriction on the smallest size of a probed sample at roughly the half-wavelength ($\lambda/2$) [26, 27]. Since THz radiation has wavelengths on the order of a millimeter, this limit impedes THz spectroscopy of microscopic samples such as high-quality exfoliated two-dimensional (2D) flakes with lateral dimensions on the order of tens of microns. Two-dimensional materials have enjoyed fantastic appeal in material and applied sciences since the exfoliation and isolation of a single layer of carbon atoms [28]. The recent observation of unconventional superconductivity in two layers of graphene precisely rotated by 1.1 degrees [29, 30] has reinvigorated investigations of correlated phenomena in these heterostructures. Near-field techniques have been developed to overcome the diffraction limit using apertures and sharp metal tips, but another promising direction is the use of on-chip waveguides and transmission lines to perform THz measurements on micrometer samples.

Sub-diffraction techniques have also revolutionized biotechnology. The zero mode waveguide presents a canonical example, now allowing researchers to perform single-molecule measurements at biologically relevant concentrations all at exceedingly small sample sizes [31, 32]. This is generally achieved by reducing the observation space to a sub-diffraction aperture of zeptoliter volume, allowing rapid and parallel measurements of single molecules at micromolar concentrations [33]. Researchers are actively trying to extend sub-diffraction techniques to the terahertz region of the electromagnetic spectrum for fingerprint analysis of bioliquids. This is challenging as water leads to strong attenuation of THz signals. For the same reason that THz radiation is appealing for identifying biomolecules, water molecules themselves present strong resonant absorptions. These absorptions mask the underlying fingerprints of the biomolecules present in a liquid sample. In Cheon et al., measurements of methylated DNA are performed at 253 K (-20°C) in order to freeze liquid water and reduce background absorption [24]. In Niessen et al. paraffin oil is used to maintain protein hydration in place of water [25]. Confining small amounts of bioliquids in microfluidic cells on the surface of on-chip THz spectrometers allows a reduction of the interaction volume between the probing field and the target sample and may provide avenues to diagnostic devices.

This chapter is organized into seven sections including this introduction. Section 2 provides a survey of the historical developments in sub-diffraction techniques for THz measurements. Section 3 presents waveguide and transmission line design considerations for on-chip measurements. Section 4 surveys research on time-domain measurements for studies of dynamics in material systems. Section 5 reviews research on THz spectroscopy on-chip and examples of spectroscopy of materials. Section 6 reviews research on THz spectroscopy of liquids. Finally, Section 7 concludes the chapter and provides an outlook on the field.

2. Historical summary of sub-diffraction techniques

Sub-diffraction techniques have been successfully used in THz imaging and microscopy, whereas sub-diffraction has proven to be more challenging [34]. Many interesting physical systems have geometries smaller than the diffraction limit: a scale less than half the wavelength of radiation where light does not interact efficiently with a material. This limit for THz radiation is on the order of 100 μm. In order to probe materials below the diffraction limit, several techniques have been developed which can be grouped into two general areas: apertures and sharp metal tips. See **Figure 1** for simple model representations of these two approaches.

Apertures form the earliest examples of sub-diffraction imaging and the idea dates back to suggestions by Synge in 1928 [37]. Aperture techniques for THz radiation were experimentally realized as early as 1993 when Merz et al. displayed an increased resolution of $\lambda/2$ [38]. Researchers have continued to improve signal resolution [39–43], up to a recent publication by Kawano, et al. where they demonstrated a 9 μm resolution using an aperture on a gallium arsenide-aluminum gallium arsenide (GaAs/AlGaAs) substrate [35]. **Figure 1**(a-b) display their device. The THz radiation is enhanced by the near-field probe and detected using the buried 2D electron gas (2DEG), as shown schematically in **Figure 1(b)**. The aperture approach is limited by the cut-off effect: as the aperture becomes smaller, and the resolution better, the transmitted signal is reduced and the signal-to-noise ratio (SNR) ultimately suffers [44, 45]. As an inconvenient technicality, the detector must be exceedingly close to the aperture to prevent signal loss from quick divergence of the radiation through the aperture.

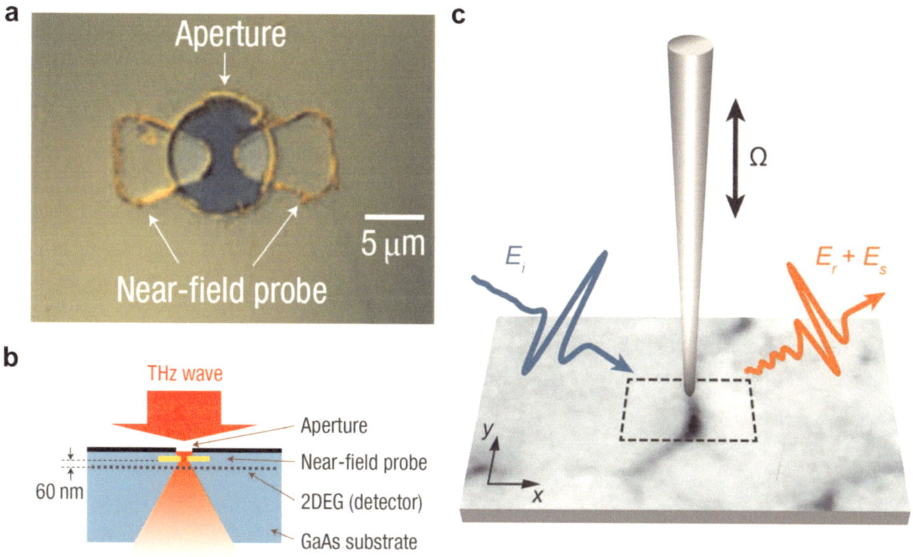

Figure 1.
Sub-diffraction terahertz imaging and spectroscopy techniques. (a) Aperture-based techniques limit the probing radiation to a small aperture, thereby increasing the resolution. This is a recent example from Kawano et al. showing an optical image of an aperture on the surface of a GaAs substrate. (b) Model schematic of the aperture device and principle of operation. The THz radiation is enhanced at the near-field probe and detected with the buried 2DEG. Panels (a) and (b) have been adapted from ref. [35]. (c) Near-field imaging and spectroscopy with a sharp metal tip. This is an example from Moon et al. showing the principle of operation. THz radiation is scattered off the metal tip and detected. Panel (c) has been adapted from ref. [36].

A second approach involves the use of a metal tip that enhances the probing field through near-field, plasmonic effects [46–49]. This was demonstrated in 1985 by Wessel using infrared radiation [46] and subsequently investigated with THz radiation [36, 47–49]. An example of this technique is shown in **Figure 1(c)**. The confinement at the apex of the metal tip produces strongly enhanced fields and sub-diffraction radiation. The increased resolution is comparable to the tip diameter (tens of nanometers). This technique has enjoyed recent interest in the imaging of 2D materials at THz frequencies [50, 51]. The technique has a few drawbacks. One, a probing field is required which may also perturb the sample under study and, two, the tip itself can lead to unwanted tip-sample interactions. Moreover, most of the scattered signal from the sample surface is background signal and the weak tip enhanced radiation requires long measurement times. Ribbek et al. have reported detection of the near-field part of the scattered signal with an integration time of 19 mins for a scan delay of 7.35 ps (amounting to a frequency resolution of 0.14 THz) [52].

Lastly, on-chip waveguides and transmission lines can be used to confine sub-diffraction radiation [53–56]. To perform spectroscopy, photoconductive switches can be fabricated near or within the on-chip circuit to generate and sample single cycle pulses using a femtosecond laser. This measurement scheme is similar to the early studies by Auston et al. that were the basis of free-space spectroscopy but that were abandoned due to pulse attenuation from lossy substrates [2]. With careful design, waveguides and transmission lines can be fabricated to achieve reflection-free spectroscopy resulting in long time-domain scans and high frequency resolution.

3. Circuit design considerations

Ultrafast pulse generation and readout on-chip have been accomplished using a number of transmission line and waveguide designs. In the earliest studies from Auston and colleagues, the chips were fabricated with microstrip lines on high-resistivity silicon [1, 57, 58]. Following this, two and three-arm coplanar transmission lines were investigated as well as coplanar waveguides [59–66]. More recently, single conductor waveguides (Goubau lines) have been used to route THz radiation and single pulses on chip [67–71]. In most cases, the circuits have been fabricated using conventional metals, but superconductors have also been used [72].

An overview of these different designs is shown in **Figure 2**. **Figure 2(a)** shows the original design by Auston [1]. The geometry consists of a continuous ground plane on the back of a silicon substrate and a hundreds-of-microns wide stripline on the surface of the chip. Auston used this design to demonstrate ultrafast (picosecond) switching of electrical signals on-chip with the aid of optical laser pulses to electrically open and close microstrip gaps. Shortly after this, Ketchen et al. measured picosecond pulses in a three arm transmission line [59]. A schematic of this design is shown in **Figure 2(b)**. Aluminum lines (5 microns wide) were patterned using standard photolithographic techniques on top of a silicon-on-sapphire wafer. The silicon was subsequently radiation-damaged to produce short (less than 1 ps) carrier lifetimes required for pulse generation and detection with a femtosecond (80 fs) laser. Using this design, the authors were able to change the relative distance between the locations of the pulse generation and detection to measure pulse dispersion. They found that the pulse width increased roughly 1 ps for a propagation length of 8 mm. **Figure 2(c)** shows an example of a coplanar waveguide design studied in ref. [70]. Here, a ground plane surrounds a center conductor and electrical lines for connections to the photoconductive switches.

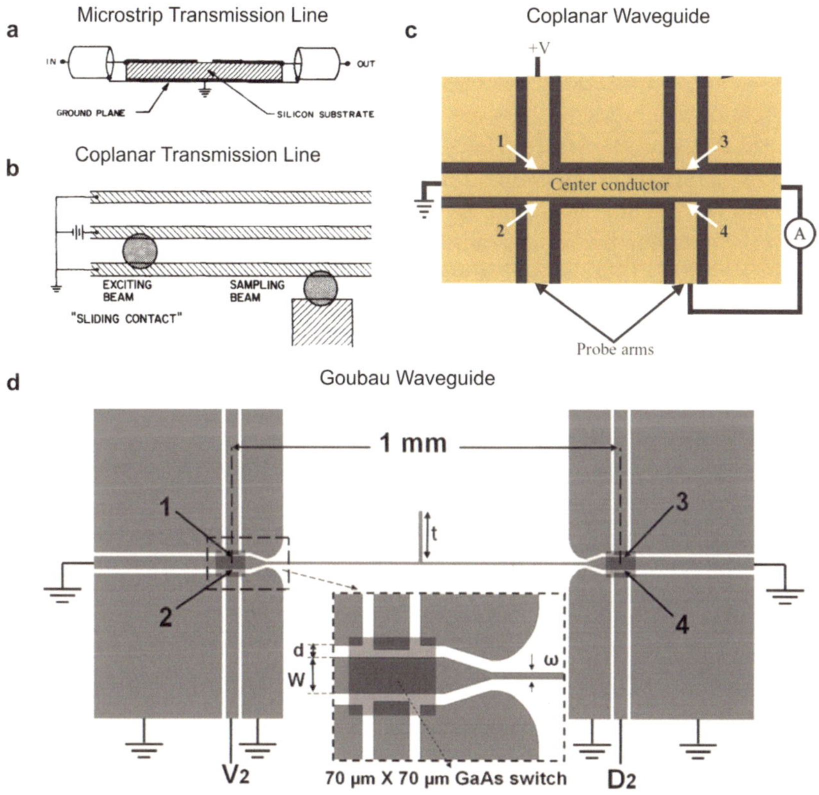

Figure 2.
Various transmission line and waveguide designs for on-chip THz studies. (a) An example of a microstrip line design. This schematic shows a side view of the device. A single conductor (hundreds of microns wide) is photolithographically patterned on top of the silicon substrate. The back side is coated with metal to function as a ground plane. This panel has been adapted from ref. [1]. (b) An example of a coplanar transmission line design. Two grounded lines surround a center conductor. Readout (sampling) is accomplished with a fourth electrical line at the bottom. This panel has been adapted from ref. [59]. (c) An example of a coplanar waveguide design. A center conductor is surrounded by ground plane patches and readout is performed with additional electrical lines (labeled probe arms). This panel has been adapted from ref. [66]. (d) An example of a Goubau (single conductor) waveguide. A single metallic strip connects two sides of the device used for pulse generation and readout. It also includes a stub filter along the transmission line of length 't' to create a band-stop filter. This panel has been adapted from ref. [70].

The circuit is created using a photolithographically defined mask and deposition of titanium (10 nm) and gold (200 nm) metals. The circuit is patterned on a gallium arsenide substrate with low temperature grown gallium arsenide (LT-GaAs) of 2 microns thickness used as photoconductive switches. The last example is a single conductor Goubau line as shown in **Figure 2(d)** from ref. [70]. Here, the single conductor from a coplanar design, such as that shown in **Figure 2(c)**, is extended across the chip without an adjacent ground plane. Dazhang et al. fabricated this structure on top of a quartz substrate using LT-GaAs as photonconductive switches.

All of these designs, in principle, can accommodate a microfluidic cell placed on top of the chip to hold small amounts of bioliquids for study. Several groups have incorporated such cells for studies of liquid analytes [73–81].

There are a few particularly attractive features of using these types of circuits fabricated on-chip. Photolithographically-defined electrodes allow control over where the radiation is routed. This includes fabricating various bends and branches to guide and even split the radiation along different paths [82]. Moreover, waveguide modes can be selectively excited. Wu et al. has shown that the dominant modes in a coplanar waveguide, slotline and coplanar modes, can be selectively excited using two photoconductive switches and a defocused laser spot [65]. This is advantageous because the field orientations for the two modes are different and one may couple better to a material or bioliquid under study. This was exactly the case in Wu et al.'s study as they found that the coplanar mode coupled better to their sample than the slotline mode. Follow up full wave modeling supported this experimental result, finding that decoupling the gate line from the ground plane better coupled the waveguide modes to the plasmon modes [83]. In such a way, the general electric field orientation can be controlled with careful choice of circuit design. For example, the coplanar transmission line design predominantly displays fields horizontal to the chip surface for materials placed between the lines. The Goubau line presents a vertical electric field for materials placed on top of the conductor. In this way, a preferential direction of the electric field excitation can be chosen.

As can be readily surmised, attenuation and pulse dispersion become significant factors in on-chip studies. As opposed to free space measurements, the pulsed radiation on-chip travels through lossy media. There are three considerations to improve the bandwidth for on-chip measurements. The first and most important is by judicious choice of a low loss substrate. Several materials have been investigated as low loss THz substrates and windows including sapphire, quartz, polyethylene, picarin, polytetrafluoroethylene (PTFE), and polyethylene terephthalate (PET) [63, 84–88]. On-chip bandwidths greater than 2 THz have been realized using picarin (Tsurupica) [63]. Another method to improve bandwidth is by thinning the substrate so the effective permittivity is lowered [89, 90]. Picarin may not support aggresive thinning (<100 μm) required for appreciable improvement but quartz is a promising alternative [89, 90]. The last method is to optimize the pulse generation and detection. The generated electric field of the THz pulse is proportional to the transient photocurrent density in the semiconducting junction [91, 92],

$$E_{THz} \propto \frac{\partial J}{\partial t} = e\left(n\frac{\partial v_n}{\partial t} + v_n\frac{\partial n}{\partial t}t + p\frac{\partial v_p}{\partial t} + v_p\frac{\partial p}{\partial t}\right), \tag{1}$$

where J is the photocurrent density, n and p are the respective electron and hole carrier densities, and v_n and v_p are the respective electron and hole average velocity. The rate of change of the densities is related to the relaxation time of the semiconducting material (τ_m) and the pulse width of the excitation laser (Δt),

$$\frac{dn}{dt} = -\frac{n}{\tau_m} + n_0 \exp - (t/\Delta t)^2, \tag{2}$$

where n_0 is the carrier density at $t = 0$ s. The same equation applies for the hole carrier density. The rate of change of the velocity is proportional to the applied electric field (E) (the voltage bias applied to the junction) through,

$$\frac{dv_{n,p}}{dt} = -\frac{v_{n,p}}{\tau_s} + \frac{q_{n,p}}{m_{n,p}}E, \tag{3}$$

where τ_s is the momentum relaxation rate and $m_{n,p}$ is the effective mass for electrons and holes respectively. Summarizing these relations, there are two aspects that will improve signal and bandwidth. The first is by using a material with a fast carrier relaxation rate and the second is utilizing a laser with a short pulse width. Radiation damaged silicon has a relaxation time of 560 fs [93] but shorter relaxation rates have been observed in LT-GaAs ($\approx$300 fs) [94].

4. Utrafast pulse measurements

Besides spectroscopy, ultrafast pulse measurements can be used to study carrier relaxation times, magnetization dynamics, and hydration processes in the time domain. Strictly speaking, these include studies that are not interested in the frequency domain and focus solely on the time-domain data.

Some of the earliest examples harness on-chip pulses to study ballistic transport in clean one-dimensional and two-dimensional systems [95, 96]. Shaner and Lyon studied two-dimensional electron gas (AlGaAs/GaAs) structures connected to an on-chip coplanar waveguide [95]. From their THz time-domain data, shown in **Figure 3(a)**, resonant oscillations appear for positive magnetic fields after a 50 ps time delay. **Figure 3(b)** displays their full data, showing the evolution of collector voltage as a function of magnetic field and time delay. The ballistic signal is distinguished by a stable signal over time (horizontal streaks), and signatures of magnetoplasmon oscillations in the 2DEG are evident by the weak signal modulation with magnetic field. Zhong et al. performed a similar experiment with carbon nanotubes connected to an on-chip coplanar transmission line [96]. They also observed ballistic carrier resonances in their time-domain data, **Figure 3(c)**, that change with the length (L_g) of the ballistic channel.

The largest category of these investigations includes emission studies of various materials including GaAs [64], InAs nanowires [98], topological insulators [99–101], silicon [102], graphene [71, 93, 97, 103, 104], molybdenum disulfide [105], carbon nanotubes [106], GaAs nanowires [107], and field emission in nanojunctions [108]. Hunter et al. demonstrated emission of picosecond pulses from graphene flakes attached to an on-chip Goubau waveguide, **Figure 3(d)** [71]. The amplitude of the pulse could be linearly controlled with the applied bias to the graphene junction (inset of **Figure 3(d)**). Using circularly polarized light, McIver et al. showed that helicity-dependent picosecond photocurrents could be measured in graphene junctions using on-chip Goubau waveguides. **Figure 3(e)** displays the data showing that the measured photocurrent changes sign when the bias direction is changed across the graphene.

Ultrafast on-chip pulses have also been used to study picosecond switching dynamics in magnets and strained semiconductors [109–116]. Of particular note, Gerrits et al. studied magnetization reversal in permalloy films (NiFe) [109]. They showed that by using two photoconductive switches to shape ultrafast pulses to excite the material, they could suppress unwanted ringing effects with a second stop pulse. Yang et al., using a coplanar transmission line, displayed magnetization reversal in GdFeCo films with 10 ps pulses [113]. The mechanism proposed for the switching is heating caused by the electrical pulses. Interestingly, only 4 fJ of energy is required to switch a 20 nm^3 magnetic cell.

5. THz spectroscopy on-chip

Using two photonconductive switches fabricated close to a transmission line or waveguide on-chip, it is possible to perform spectroscopy from the time-domain data

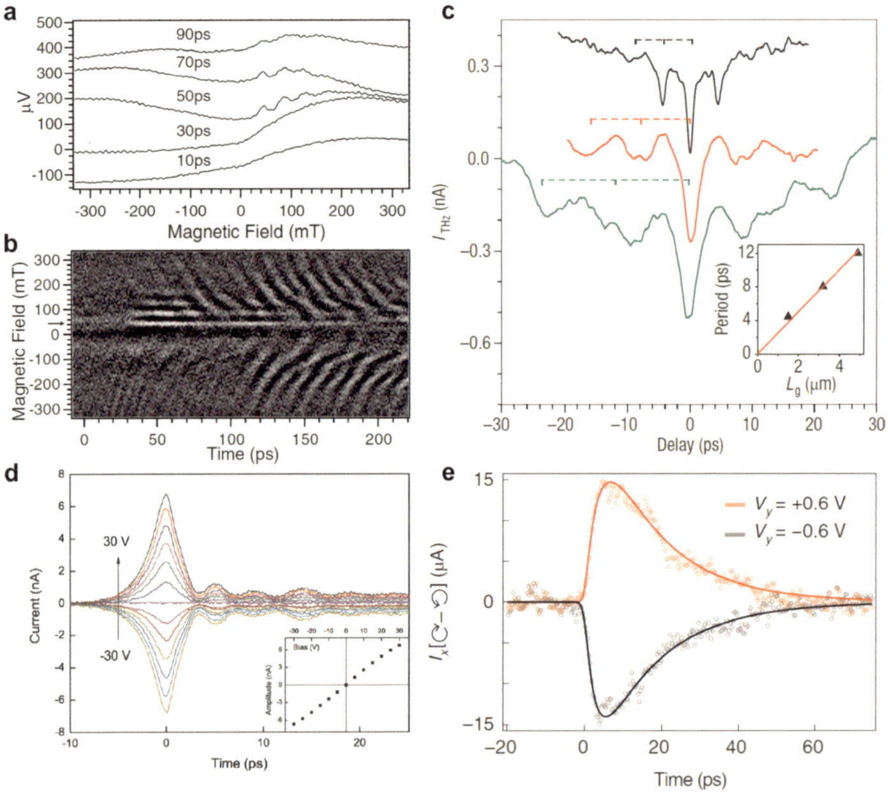

Figure 3.
Time-domain studies of various materials. (a) Ballistic carrier transport in a two-dimensional electron gas (2DEG). The collector voltage (y-axis) is plotted as a function of perpendicular magnetic field for different time delays between the generation switch and collector. Oscillations corresponding to ballistic transport can be seen for delays of 50 ps and greater. (b) Collector voltage plotted as a function of magnetic field and time delay. Horizontal streaks in the data correspond to ballistic transport and weaker oscillations are from magnetoplasmon excitations. Panels (a) and (b) have been adapted from ref. [95]. (c) Drain current of a carbon nanotube device as a function of time delay between two on-chip THz pulses. Dips in the data correspond to ballistic carrier transport. The curves from top to bottom correspond to ballistic channels with lengths of L_g = 1.5 mm (black), 3.2 mm (red) and 4.9 mm (green). This panel has been adapted from ref. [96]. (d) Readout current as a function of time delay between generation and detection pulses for a graphene junction emission switch. The various curves correspond to different voltage biases applied to the graphene flake. This panel has been adapted from ref. [71]. (e) Helicity-dependent photocurrent as a function of time delay between excitation pulse and readout pulse. The two curves show that the photocurrent changes sign for different orientations of the applied voltage bias, as expected for anomalous hall currents. This panel has been adapted from ref. [97].

using either pulsed lasers or continuous wave optical beating. Several research groups have investigated circuit elements that lead to resonant features in the THz spectrum. For example, Dazhang et al. have used a Goubau waveguide with a stub line, shown in **Figure 2(d)**, to create a band-stop filter [70]. **Figure 4(a)** displays their data showing the measured time-domain scan in the inset and the Fourier transformed data in the main panel. Band-stop response can be seen at the location of the arrows centered at 250 GHz (fundamental mode) and 780 GHz (third harmonic).

Very recently, Yoshioka et al. have demonstrated on-chip spectroscopy using continuous wave THz radiation created from the optical beating of two near-IR lasers [117]. This technique offers greater frequency resolution (10 MGz vs. 10 Ghz) when compared with the time-domain technique. **Figure 4(b)** displays data from their

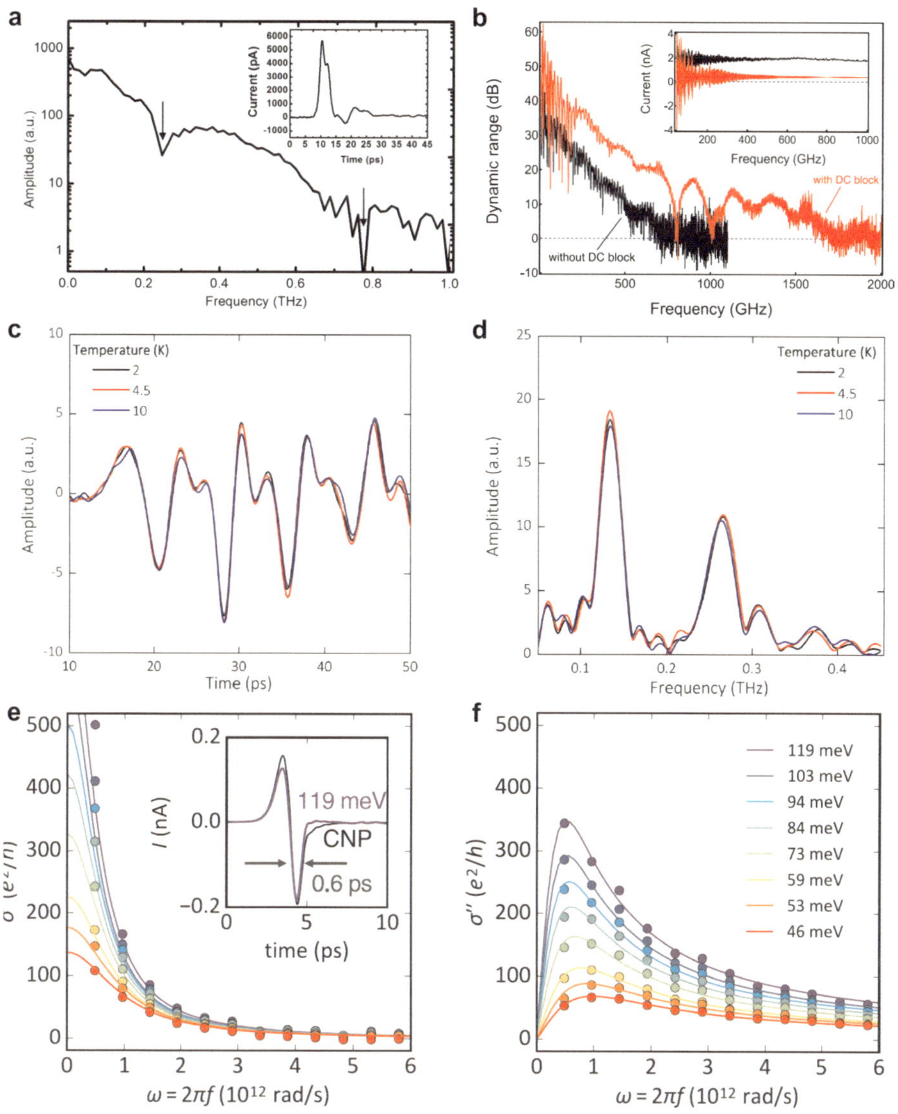

Figure 4.
Examples of on-chip THz spectroscopy. (a) Transmission spectra of a Goubau waveguide device with a band-stop filter (a stub line fabricated on the Goubau line shown in ***Figure 2(d)****). The dips with arrows indicate the band stop frequencies of 250 GHz (fundamental mode) and 780 GHz (third harmonic). The inset shows the time domain scan used to form the spectra in the main panel from a Fourier transform. This panel has been adapted from ref. [70]. (b) Amplitude spectra obtained for a coplanar waveguide device with and without a DC block (discontinuity in the central conductor). The inset shows the raw current measurement from which the amplitude spectra is calculated. This panel has been adapted from ref. [117]. (c) Time-domain data at different temperatures of a coplanar device with a 2DEG (AlGaAg/GaAs) mesa integrated within the centerline conductor. (d) Amplitude spectra of the Fourier transformed time-domain data in panel (c). Resonant peaks correspond to plasmon excitations in the gated (132 GHz and 264 GHz) and ungated (311 GHz) regions. Panels (c) and (d) have been adapted from ref. [118]. (e) Real part of the complex conductivity of a graphene heterostructure integrated within a coplanar transmission line device. The colored curves correspond to different Fermi energies labeled in the legend in panel (f). The solid lines are fits to the Drude model. The inset shows the time-domain data for two different Fermi energies. (f) The imaginary part of the complex conductivity of a graphene heterostructure. The solid lines are fits to the Drude model. Panels (e) and (f) have been adapted from ref. [119].*

work. They investigated devices with and without a DC block (a gap in the centerline conductor) in a coplanar waveguide structure. The data (red) measured with a DC block show dips in the spectra due to interference of even and odd modes of the waveguide. The DC block enhances the bandwidth spectrometer and reduces unwanted DC leakage currents.

On-chip THz spectrometers have also been used to study plasmon excitations in clean 2DEG systems [118, 120, 121]. Wu et al. showed that plasmon excitations in an etched mesa of AlGaAs/GaAs could be electrically gated [118]. **Figure 4(c)** shows their time-domain data. Very prominent oscillations with different periods can be seen. The Fourier transform of the time-domain data is shown in **Figure 4(d)**. Three peaks are produced at frequencies of 132, 264, and 311 GHz. The lower frequencies coincide with the predicted plasmon frequency of the first and second mode for the gated region of the device. The third resonance agrees with the plasmon frequency expected for the ungated region.

One of the qualities of the on-chip system, as highlighted in the introduction, is the ability to probe materials below the diffraction limit. Clean exfoliated 2D materials are typically tens of microns on a side and are well-suited for investigation with on-chip spectroscopy. Gallagher et al. have probed the THz properties of a graphene heterostructure using a coplanar transmission line and LT-GaAs photoconductive switches [119]. They investigated the predicted quantum critical characteristics of graphene at low doping and its hydrodynamic properties at higher doping. The inset of **Figure 4(e)** displays their time-domain data at two different dopings. The main panels of **Figure 4(e-f)** show the optical conductivity (real and imaginary, respectively) calculated from the Fourier transform of the time-domain data for various Fermi energies corresponding to the legend shown in **Figure 4(f)**. The data follow the Drude formula closely, $\sigma = D_{gr}\pi^{-1}(\tau^{-1} - i\omega)^{-1}$ with the known Drude weight of graphene, $D_{gr} = 2(e/\hbar)^2 k_B T_e \log\left(2\cosh\left(m(2k_B T_e)^{-1}\right)\right)$ and τ the only fitting parameter. This data shows the Fermi liquid character of graphene at finite doping. The authors go on to extract the scattering rates for charge neutral graphene, finding support for quantum critical behavior in their linear dependence with temperature.

6. THz spectroscopy of liquids

In addition to solids, several research groups have incorporated reservoirs and microfluidic channels to perform spectroscopy on liquids. This could open up an avenue toward diagnostic devices for the detection of low-concentration biomolecules or investigations of chemical reactions *in-situ*.

Probably the most straightforward approach to investigating liquids is by fabricating a microfluidic channel or reservoir on the surface of the chip [73–81]. Swithenbank et al. has shown that a polydimethylsiloxane (PDMS) molded microfluidic channel can be plasma-bonded to the surface of a Goubau waveguide chip [74]. **Figure 5(a)** shows a model schematic of the device. A thin (6 μm) layer of benzocyclobutene is spun on the chip before the PDMS mold to electrically isolate the Goubau line from the liquid channel. Using this device, they measured the complex permittivity of several liquids in the THz regime including methanol, ethanol, propan-1-ol, butan-1-ol, hexan-1-ol, octan-1-ol, and mixtures of propan-2-ol/DI-H2O. **Figure 5(b)** displays measurements on mixtures of propan-2-ol/DI-H2O for different concentrations of DI-H2O in 10% increments from red (0%) to purple (100%). The

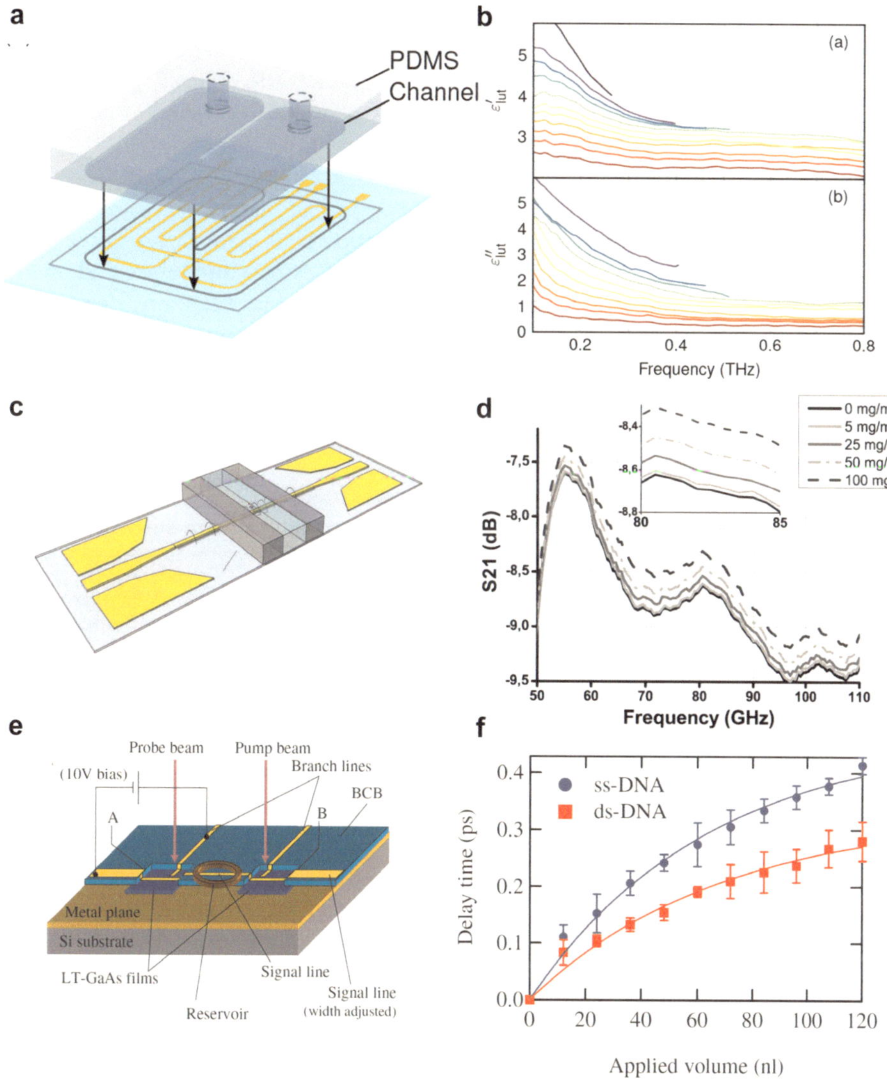

Figure 5.
Examples of on-chip THz spectroscopy of liquids. (a) Model schematic of an on-chip spectrometer for measurements of liquids. A PDMS mold is plasma bonded to the surface of a Goubau line device. (b) Complex permittivity measured for various mixtures of propan-2-ol/DI-H2O for different concentrations of DI-H2O in 10% increments from red (0%) to purple (100%). Panels (a) and (b) have been adapted from ref. [74]. (c) Model schematic of an on-chip spectrometer for measurements of liquids. A silicon microfluidic channel is fabricated on the surface of a Goubau line device. (d) Transmission (S21) measurements of various concentrations of BSA in DI water. Panels (c) and (d) have been adapted from ref. [75]. (e) Model schematic of an on-chip spectrometer for measurements of liquids. A reservoir made from SU-8 photoresist is created on the surface of a chip with a microstrip line. (f) THz pulse delay as a function of sample volumes of single-stranded (ss) and double-stranded DNA. The solid lines are fits to an exponential model. Panels (e) and (f) have been adapted from ref. [80].

complex permittivity can be seen to gradually increase with the percentage of DI-H2O, as expected. One can also appreciate from the data that the bandwidth of the measurement decreases as the mixture becomes more absorbing. This highlights a design parameter of the spectrometer that is important. If greater bandwidth for a relatively absorbing substance is required, one can reduce the interaction volume between the probing field and the measured liquid by shrinking the volume of the microfluidic channel.

Another example of this sort of device is given by Laurette et al. [75]. They also used a Goubau line device but instead of using photoconductive switches pumped with femtosecond lasers, they connected the chip to a high-frequency vector network analyzer with frequency bands of 0–110 GHz and 140–220 GHz. They measured mixtures of bovine Serum Albumin (BSA) in DI water and lysozyme in DI water at different concentrations to study hydration shell structure. **Figure 5(c)** shows a model schematic of the device consisting of a Goubau line fabricated on a pyrex borosilicate glass substrate. The microfluidic channel is created by deep reactive ion etching of silicon that is bonded to the glass substrate [81]. **Figure 5(d)** shows the measured transmission (S21) for various concentrations of BSA powders. The transmission amplitude is seen to increase with BSA concentration. They found a lower detectable sensitivity limit of the system of 5 mg/mL^2 for protein spectroscopy.

Kasai et al. have demonstrated THz pulse measurements of DNA solutions using an on-chip device consisting of a microstrip line fabricated on a silicon substrate, **Figure 5(e)** [80]. A sample reservoir with a diameter of 400 μm and a height of 6.5 μm was created from SU-8 photoresist. **Figure 5(f)** displays their results showing that the pulse delay tracks with the amount of single-stranded (ss) and double-stranded (ds) DNA samples. The pulse's propagation is exponentially delayed with increased sample volume and follows the expected trend of $t = t_{sat}[1 - \exp(-d/V)]$, where V is the sample volume, and t_{sat} and d are fitting parameters. The distinct difference between the single-stranded and double-stranded results paves the way for a diagnostic device for biomedical sciences.

Finally, highlighting the versatility of on-chip spectroscopy, Russel et al. [122] have presented measurements of various alcohols using a Goubau line fabricated on a flexible polyimide substrate wrapped around a quartz capillary (3 mm diameter with 10 μm thick walls). Using this system they were able to measure the time-of-flight permittivities [123] of various alcohols. The measurement only provided relative approximations of the permittivities but the signal-to-noise ratio was found to be better than that obtained from free-space measurements of the same alcohols. The flexible substrate approach provides another degree-of-freedom in the design of on-chip systems.

7. Conclusions

This chapter has reviewed the history and common applications of on-chip THz time-domain spectroscopy. Free-space THz methods struggle to probe samples smaller than the diffraction limit. Still, several techniques have been developed to overcome this limit using apertures, sharp metal tips, and on-chip waveguides. Common designs include coplanar transmission lines and Goubau waveguides. Using photolithography and metallization, the circuit design has many freedoms, useful for signal routing, filtering, and even selecting the propagation mode of the electric field.

Using this technique, researchers have investigated electronic properties of materials in both the time and frequency domain. Time-domain data can show the evolution of a system with picosecond resolution. This has been used to show rapid phenomena, such as magnetoplasmon oscillations in GaAs/AlGaAs 2DEGs, ballistic carrier resonances in carbon nanotubes, and emission and relaxation characteristics of a variety of photoconductive materials impinged by a femtosecond laser pulse. Frequency-domain data is calculated using a simple discrete Fourier transform of the time domain. This absorption spectra have been used to characterize on-chip band filters, assess the efficacy of DC blocks in the transmission line, and directly measure plasmon excitation frequencies in gated GaAs/AlGaAs. Additionally, the frequency-dependent complex conductivity of materials can be extracted from the THz spectra, enabling deep analysis of the scattering rates of electronic systems.

The modularity of chips allows microscopic liquid samples to be probed as well. Various designs for microfluidic cells have been developed, including reservoirs made of PDMS and photoresist, and channels etched into a silicon substrate, each fabricated atop a transmission line. These developments are especially useful for biomedical applications, as minimal sample size is necessary for comprehensive results. Studies include characterization of frequency bandwidth with various volumes of sample, determination of sensitivity to small samples of protein, and distinction between single-stranded and double-stranded DNA samples. The Goubau line design is also capable of fitting on a flexible substrate, demonstrating its versatility.

Terahertz spectroscopy has great potential in probing a realm of physics inaccessible by other techniques and applying it on a small-footprint chip opens up many interesting directions for future research. Spectroscopy on nanomaterials is one clear avenue for further development. THz spectral information could lead to a better understanding of complex ground states in 2D materials and heterostructures. One example is the recently discovered twisted heterostructures that present superconductivity such as twisted bilayer graphene. On-chip THz spectroscopy offers a method to probe these types of samples that are too small for free space measurements. The transmission spectra acquired could provide detailed tracking of spectral weight and sensitive investigation of the onset of interactions and correlations in these systems. Moreover, using on-chip spectroscopy with an optical pump would enable studies of quasiparticle dynamics and information on possible pairing mechanisms that are not well understood in twisted heterostructures.

Given the progress so far, another clear direction of development is the study of bioliquid samples on-chip. Many biomolecular motions present vibrational and rotational excitations in the THz spectrum. These excitations can serve as spectral fingerprints of the species for in-vitro diagnosis. Special attention will be needed on signal attenuation from water but careful adjustment of the interaction volume may lead to the detection of very low-concentration biomolecules in liquids. With advances in compact THz sources and detectors, new diagnostic devices with better sensitivity may be developed.

Acknowledgements

This material is based upon work supported by the National Science Foundation under Grant No. (2047509).

Conflict of interest

The authors declare no conflict of interest.

Author details

Randy M. Sterbentz and Joshua O. Island*
Department of Physics and Astronomy, University of Nevada Las Vegas, Las Vegas, USA

*Address all correspondence to: joshua.island@unlv.edu

IntechOpen

© 2023 The Author(s). Licensee IntechOpen. This chapter is distributed under the terms of the Creative Commons Attribution License (http://creativecommons.org/licenses/by/3.0), which permits unrestricted use, distribution, and reproduction in any medium, provided the original work is properly cited.

References

[1] Auston DH. Picosecond optoelectronic switching and gating in silicon. Applied Physics Letters. 1975; **26**(3):101-103

[2] Auston DH, Cheung KP, Smith PR. Picosecond photoconducting Hertzian dipoles. Applied Physics Letters. 1984; **45**(3):284-286

[3] Smith PR, Auston DH, Nuss MC. Subpicosecond photoconducting dipole antennas. IEEE Journal of Quantum Electronics. 1988;**24**(2):255-260

[4] Fattinger C, Grischkowsky D. Point source terahertz optics. Applied Physics Letters. 1988;**53**(16):1480-1482

[5] Fattinger C, Grischkowsky D. Terahertz beams. Applied Physics Letters. 1989;**54**(6):490-492

[6] van Exter M, Fattinger C, Grischkowsky D. Terahertz time-domain spectroscopy of water vapor. Optics Letters. 1989;**14**(20):1128-1130

[7] Huber R, Tauser F, Brodschelm A, Bichler M, Abstreiter G, Leitenstorfer A. How many-particle interactions develop after ultrafast excitation of an electron–hole plasma. Nature. 2001;**414**(6861): 286-289

[8] Dexheimer S, L. Terahertz Spectroscopy: Principles and Applications. Boca Raton: CRC Press; 2017

[9] Porer M, Leierseder U, Ménard J-M, Dachraoui H, Mouchliadis L, Perakis IE, et al. Non-thermal separation of electronic and structural orders in a persisting charge density wave. Nature Materials. 2014;**13**(9):857-861

[10] Hilton DJ, Prasankumar RP, Trugman SA, Taylor AJ, Averitt RD. On photo-induced phenomena in complex materials: Probing quasiparticle dynamics using infrared and far-infrared pulses. Journal of the Physical Society of Japan. 2006;**75**(1):011006

[11] Averitt RD, Lobad AI, Kwon C, Trugman SA, Thorsmølle VK, Taylor AJ. Ultrafast conductivity dynamics in colossal magnetoresistance manganites. Physical Review Letters. 2001;**87**(1): 017401

[12] Aoki H. Novel Landau level laser in the quantum hall regime. Applied Physics Letters. 1986;**48**(9):559-560

[13] Morimoto T, Hatsugai Y, Aoki H. Cyclotron radiation and emission in graphene. Physical Review B. 2008; **78**(7):073406

[14] Wendler F, Malic E. Towards a tunable graphene-based Landau level laser in the terahertz regime. Scientific Reports. 2015;**5**(1):12646

[15] Yang X, Zhao X, Yang K, Yueping Liu Y, Liu WF, Luo Y. Biomedical applications of terahertz spectroscopy and imaging. Trends in Biotechnology. 2016;**34**(10):810-824

[16] Jing X, Plaxco KW, James S, Allen. Probing the collective vibrational dynamics of a protein in liquid water by terahertz absorption spectroscopy protein. Science. 2006;**15**(5):1175-1181

[17] Dinca MP, Leca A, Apostol D, Mernea M, Calborean O, Mihailescu D, et al. Transmission thz time domain system for biomolecules spectroscopy. Journal of Optoelectronics and Advanced Materials. 2010;**12**(January 2010):110-114

[18] Sczech R, Haring P, Bolvar. Thz spectroscopy of bovine serum albumin

solution using the long-range guided mode supported by thin liquid films. In: 2014 Conference on Lasers and Electro-Optics (CLEO)-Laser Science to Photonic Applications. San Jose, CA, USA: IEEE; 2014. pp. 1-2

[19] Yoon SA, Cha SH, Jun SW, Park SJ, Park J-Y, Lee S, et al. Identifying different types of microorganisms with terahertz spectroscopy. Biomedical Optics Express. 2020;**11**(1):406-416

[20] Fan S, He Y, Ung BS, Pickwell-MacPherson E. The growth of biomedical terahertz research. Journal of Physics D: Applied Physics. 2014;**47**(37): 374009

[21] Pickwell E, Cole BE, Fitzgerald AJ, Pepper M, Wallace VP. In vivo study of human skin using pulsed terahertz radiation. Physics in Medicine & Biology. 2004;**49**(9):1595

[22] Wallace VP, Fitzgerald AJ, Shankar S, Flanagan N, Pye R, Cluff J, et al. Terahertz pulsed imaging of basal cell carcinoma ex vivo and in vivo. British Journal of Dermatology. 2004; **151**(2):424-432

[23] Yao X, Havenith M. Perspective: Watching low-frequency vibrations of water in biomolecular recognition by thz spectroscopy. The Journal of Chemical Physics. 2015;**143**(17):170901

[24] Cheon H, Yang H-j, Lee S-H, Kim YA, Son J-H. Terahertz molecular resonance of cancer DNA. Scientific Reports. 2016;**6**(1):1-10

[25] Niessen KA, Mengyang X, George DK, Chen MC, Adrian R, D'Amaré F, et al. Protein and RNA dynamical fingerprinting. Nature Communications. 2019;**10**(1): 1-10

[26] Abbe E. Beiträge zur Theorie des Mikroskops und der mikroskopischen Wahrnehmung. Archiv für Mikroskopische Anatomie. 1873;**9**(1): 413-468

[27] Lord Rayleigh FRSXXXI. Investigations in optics, with special reference to the spectroscope. The London, Edinburgh, and Dublin Philosophical Magazine and Journal of Science. 1879;**8**(49):261-274

[28] Novoselov KS, Geim AK, Morozov SV, Jiang D, Zhang Y, Dubonos SV, et al. Electric field effect in atomically thin carbon films. Science. 2004;**306**(5696):666-669

[29] Cao Y, Fatemi V, Fang S, Watanabe K, Taniguchi T, Kaxiras E, et al. Unconventional superconductivity in magic-angle graphene superlattices. Nature. 2018;**556**(7699):43-50

[30] Cao Y, Fatemi V, Demir A, Fang S, Tomarken SL, Luo JY, et al. Correlated insulator behaviour at half-filling in magic-angle graphene superlattices. Nature. 2018;**556**(7699):80-84

[31] Levene MJ, Korlach J, Turner SW, Foquet M, Craighead HG, Webb WW. Zero-mode waveguides for single-molecule analysis at high concentrations. Science. 2003;**299**(5607):682-686

[32] Moran-Mirabal JM, Craighead HG. Zero-mode waveguides: Sub-wavelength nanostructures for single molecule studies at high concentrations. Methods. 2008;**46**(1):11-17

[33] Zhu P, Craighead HG. Zero-mode waveguides for single-molecule analysis. Annual Review of Biophysics. 2012;**41**: 269-293

[34] Adam AJL. Review of near-field terahertz measurement methods and

their applications. Journal of Infrared, Millimeter, and Terahertz Waves. 2011; **32**(8):976

[35] Kawano Y, Ishibashi K. An on-chip near-field terahertz probe and detector. Nature Photonics. 2008;**2**(10):618-621

[36] Moon K, Do Y, Park H, Kim J, Kang H, Lee G, et al. Computed terahertz near-field mapping of molecular resonances of lactose stereo-isomer impurities with sub-attomole sensitivity. Scientific Reports. 2019;**9**(1): 1-8

[37] Synge EH, XXXVIII. A suggested method for extending microscopic resolution into the ultra-microscopic region. The London, Edinburgh, and Dublin Philosophical Magazine and Journal of Science. 1928;**6**(35):356-362

[38] Merz R, Keilmann F, Haug RJ, Ploog K. Nonequilibrium edge-state transport resolved by far-infrared microscopy. Physical Review Letters. 1993;**70**(5):651-653

[39] Hunsche S, Koch M, Brener I, Nuss MC. THz near-field imaging. Optics Communications. 1998;**150**(1): 22-26

[40] Nguyen TD, Valy Vardeny Z, Nahata A. Concentration of terahertz radiation through a conically tapered aperture. Optics Express. 2010;**18**(24): 25441-25448

[41] Rusina A, Durach M, Nelson KA, Stockman MI. Nanoconcentration of terahertz radiation in plasmonic waveguides. Optics Express. 2008; **16**(23):18576-18589

[42] Mitrofanov O, Brener I, Harel R, Wynn JD, Pfeiffer LN, West KW, et al. Terahertz near-field microscopy based on a collection mode detector. Applied Physics Letters. 2000;**77**(22): 3496-3498

[43] Carnio BN, Elezzabi AY. Investigation of ultra-broadband terahertz generation from sub-wavelength lithium niobate waveguides excited by few-cycle femtosecond laser pulses. Optics Express. 2017;**25**(17): 20573-20583

[44] Bethe HA. Theory of diffraction by small holes. Physical Review. 1944;**66**(7–8):163-182

[45] Bouwkamp CJ. On the diffraction of electromagnetic waves by small circular disks and holes. Philips Research Reports. 1950;**5**:401-422

[46] Wessel J. Surface-enhanced optical microscopy. JOSA B. 1985;**2**(9): 1538-1541

[47] van der Valk NCJ, Planken PCM. Electro-optic detection of subwavelength terahertz spot sizes in the near field of a metal tip. Applied Physics Letters. 2002;**81**(9):1558-1560

[48] Chen H-T, Kersting R, Cho GC. Terahertz imaging with nanometer resolution. Applied Physics Letters. 2003;**83**(15):3009-3011

[49] Huber AJ, Keilmann F, Wittborn J, Aizpurua J, Hillenbrand R. Terahertz near-field Nanoscopy of Mobile carriers in single semiconductor Nanodevices. Nano Letters. 2008;**8**(11):3766-3770

[50] Zhang J, Chen X, Mills S, Ciavatti T, Yao Z, Mescall R, et al. Terahertz nanoimaging of graphene. ACS Photonics. 2018;**5**(7):2645-2651

[51] Yao Z, Semenenko V, Zhang J, Mills S, Zhao X, Chen X, et al. Photo-induced terahertz near-field dynamics of

graphene/InAs heterostructures. Optics Express. 2019;**27**(10):13611-13623

[52] von Ribbeck H-G, Brehm M, van der Weide DW, Winnerl S, Drachenko O, Helm M, et al. Spectroscopic THz near-field microscope. Optics Express. 2008; **16**(5):3430-3438

[53] Sprik R, Duling IN, Chi C-C, Grischkowsky D. Far infrared spectroscopy with subpicosecond electrical pulses on transmission lines. Applied Physics Letters. 1987;**51**(7): 548-550

[54] Matheisen C, Nagel M, Sawallich S. Coplanar stripline (CPS) emitter and transceiver microprobes for ultra-high bandwidth on-chip terahertz measurements. In: 2015 40th International Conference on Infrared, Millimeter, and Terahertz Waves (IRMMW-THz). Hong Kong, China: IEEE; 2015. pp. 1-3

[55] Saxler J, Gómez Rivas J, Janke C, Pellemans HPM, Haring Bolívar P, Kurz H. Time-domain measurements of surface plasmon polaritons in the terahertz frequency range. Physical Review B. 2004;**69**(15): 155427

[56] O'Hara JF, Averitt RD, Taylor AJ. Prism coupling to terahertz surface plasmon polaritons. Optics Express. 2005;**13**(16):6117-6126

[57] Auston DH, Johnson AM, Smith PR, Bean JC. Picosecond optoelectronic detection, sampling, and correlation measurements in amorphous semiconductors. Applied Physics Letters. 1980;**37**(4):371-373

[58] Auston D. Impulse response of photoconductors in transmission lines. IEEE Journal of Quantum Electronics. 1983;**19**(4):639-648

[59] Ketchen MB, Grischkowsky D, Chen TC, Chi C-C, Duling Iii IN, Halas NJ, et al. Generation of subpicosecond electrical pulses on coplanar transmission lines. Applied Physics Letters. 1986;**48**(12):751-753

[60] Paulter NG, Sinha DN, Gibbs AJ, Eisenstadt WR. Optoelectronic measurements of picosecond electrical pulse propagation in coplanar waveguide transmission lines. IEEE Transactions on Microwave Theory and Techniques. 1989;**37**(10):1612-1619

[61] Grischkowsky DR, Ketchen MB, Chi C-C, Duling IN, Halas NJ, Halbout J-M, et al. Capacitance free generation and detection of subpicosecond electrical pulses on coplanar transmission lines. IEEE Journal of Quantum Electronics. 1988;**24**(2):221-225

[62] Jacobsen RH, Karen Birkelund T, Holst PU, Jepsen, and SR Keiding. Interpretation of photocurrent correlation measurements used for ultrafast photoconductive switch characterization. Journal of Applied Physics. 1996;**79**(5):2649-2657

[63] Yanagi S, Onuma M, Kitagawa J, Kadoya Y. Propagation of terahertz pulses on coplanar strip-lines on low permittivity substrates and a spectroscopy application. Applied Physics Express. 2008;**1**(1):012009

[64] Prechtel L, Manus S, Schuh D, Wegscheider W, Holleitner AW. Spatially resolved ultrafast transport current in gaas photoswitches. Applied Physics Letters. 2010;**96**(26):261110

[65] Jingbo W, Mayorov AS, Wood CD, Mistry D, Li L, Muchenje W, et al. Excitation, detection and electrostatic manipulation of terahertz-frequency range plasmons in a two-dimensional

electron system. Scientific Reports. 2015; **5**(1):1-8

[66] Wood CD, Mistry D, Li LH, Cunningham JE, Linfield EH, Davies AG. On-chip terahertz spectroscopic techniques for measuring mesoscopic quantum systems. Review of Scientific Instruments. 2013;**84**(8): 085101

[67] Treizebré A, Bocquet B, Yansheng X, Bosisio RG. New thz excitation of planar goubau line. Microwave and Optical Technology Letters. 2008;**50**(11): 2998-3001

[68] Gacemi D, Mangeney J, Laurtent T, Lampin J-F, Akalin T, Blary K, et al. Thz surface plasmon modes on planar goubau lines. Optics Express. 2012; **20**(8):8466-8471

[69] Zehar M, Moreno G, Chahadih A, Turer I, Ghaddar A, Akalin T. Low loss terahertz planar goubau line on high resistivity silicon substrate. In: 2013 13th Mediterranean Microwave Symposium (MMS). Saida, Lebanon: IEEE; 2013. pp. 1-3

[70] Dazhang L, Cunningham J, Byrne MB, Khanna S, Wood CD, Burnett AD, et al. On-chip terahertz goubau-line waveguides with integrated photoconductive emitters and mode-discriminating detectors. Applied Physics Letters. 2009;**95**(9):092903

[71] Hunter N, Mayorov AS, Wood CD, Russell C, Li L, Linfield EH, et al. On-Chip picosecond pulse detection and generation using graphene photoconductive switches. Nano Letters. 2015;**15**(3):1591-1596

[72] Chi C, Gallagher W, Duling I, Grischkowsky D, Halas N, Ketchen M, et al. Subpicosecond optoelectronic study of superconducting transmission lines. IEEE Transactions on Magnetics. 1987;**23**(2):1666-1669

[73] Amarloo H, Safavi-Naeini S. Enhanced on-chip terahertz vibrational absorption spectroscopy using evanescent fields in silicon waveguide structures. Optics Express. 2021;**29**(11): 17343-17352

[74] Swithenbank M, Burnett AD, Russell C, Li LH, Davies AG, Linfield EH, et al. On-chip terahertz-frequency measurements of liquids. Analytical Chemistry. 2017;**89**(15): 7981-7987

[75] Laurette S, Treizebre A, Elagli A, Hatirnaz B, Froidevaux R, Affouard F, et al. Highly sensitive terahertz spectroscopy in microsystem. RSC Advances. 2012;**2**(26):10064-10071

[76] Park SJ, Cunningham J. Determination of permittivity of dielectric analytes in the terahertz frequency range using split ring resonator elements integrated with on-chip waveguide. Sensors. 2020;**20**(15): 4264

[77] Tonouchi M. Terahertz microfluidic chip sensitivity-enhanced with a few arrays of meta atoms. In: Optical Sensors. Vancouver, British Columbia, Canada: Optica Publishing Group; 2018. p. SeTh4E–1

[78] Serita K, Kobatake S, Tonouchi M. I-design terahertz microfluidic chip for attomolar-level sensing. Journal of Physics: Photonics. 2022;**4**(3):034005

[79] Swithenbank M, Russell C, Burnett AD, Li LH, Linfield H, Davies G, et al. On-chip terahertz spectroscopy of liquid mixtures. In: 2015 40th International Conference on Infrared, Millimeter, and Terahertz Waves

(IRMMW-THz). Hong Kong, China: IEEE; 2015. pp. 1-3

[80] Kasai S, Tanabashi A, Kajiki K, Itsuji T, Kurosaka R, Yoneyama H, et al. Micro strip line-based on-chip terahertz integrated devices for high sensitivity biosensors. Applied Physics Express. 2009;**2**(6):062401

[81] Laurette S, Treizebre A, Bocquet B. Co-integrated microfluidic and THz functions for biochip devices. Journal of Micromechanics and Microengineering. 2011;**21**(6):065029

[82] Sato N, Kitagawa J, Kadoya Y. THz pulse propagation on microstrip discontinuities. Journal of Infrared, Millimeter, and Terahertz Waves. 2011; **32**(5):666-672

[83] Dawood A, Wu JB, Wood CD, Li LH, Linfield EH, Davies AG, et al. Full-wave modelling of terahertz frequency plasmons in two-dimensional electron systems. Journal of Physics D: Applied Physics. 2019;**52**(21):215101

[84] Kadoya Y. THz wave propagation on strip lines: Devices, properties, and applications. In: 2007 19th International Conference on Applied Electromagnetics and Communications. Dubrovnik, Croatia: IEEE; 2007. pp. 1-4

[85] Ohta N, Yanagi S, Onuma M, Kitagawa J, Kadoya Y. Propagation of terahertz pulses on polymer-based coplanar striplines. In: 2008 Asia-Pacific Microwave Conference. Hong Kong, China: IEEE; 2008. pp. 1-4

[86] Hejase JA, Paladhi PR, Chahal PP. Terahertz characterization of dielectric substrates for component design and nondestructive evaluation of packages. IEEE Transactions on Components, Packaging and Manufacturing Technology. 2011;**1**(11):1685-1694

[87] Song H-J. Packages for terahertz electronics. Proceedings of the IEEE. 2017;**105**(6):1121-1138

[88] Fernandez N, Zimmermann P, Zechmann P, Wörle M, Kienberger R, Holleitner A. Toward femtosecond electronics up to 10 THz. In: Ultrafast Phenomena and Nanophotonics XXIII. Vol. 10916. International Society for Optics and Photonics; 2019. p. 109160R

[89] Russell C, Wood CD, Dazhang L, Burnett AD, Li LH, Linfield EH, et al. Increasing the bandwidth of planar on-chip THz devices for spectroscopic applications. In: 2011 International Conference on Infrared, Millimeter, and Terahertz Waves. Houston, TX, USA. 2011. pp. 1-3

[90] Russell C, Wood CD, Burnett AD, Li L, Linfield EH, Giles Davies A, et al. Spectroscopy of polycrystalline materials using thinned-substrate planar Goubau line at cryogenic temperatures. Lab on a Chip. London, United Kingdom: Royal Society of Chemistry; 2013;**13**(20): 4065-4070

[91] Piao Z, Tani M, Sakai K. Carrier dynamics and terahertz radiation in photoconductive antennas. Japanese Journal of Applied Physics. 2000;**39**(1R): 96

[92] Russell C. Broadband on-chip terahertz spectroscopy. [Phd thesis] University of Leeds. 2013.

[93] Island JO, Kissin P, Schalch J, Cui X, Haque SRU, Potts A, et al. On-chip terahertz modulation and emission with integrated graphene junctions. Applied Physics Letters. 2020;**116**(16):161104

[94] Gupta S, Whitaker JF, Mourou GA. Ultrafast carrier dynamics in III-V semiconductors grown by molecular-beam epitaxy at very low substrate

temperatures. IEEE Journal of Quantum Electronics. 1992;**28**(10):2464-2472

[95] Shaner EA, Lyon SA. Picosecond time-resolved two-dimensional ballistic electron transport. Physical Review Letters. 2004;**93**(3):037402

[96] Zhong Z, Gabor NM, Sharping JE, Gaeta AL, McEuen PL. Terahertz time-domain measurement of ballistic electron resonance in a single-walled carbon nanotube. Nature Nanotechnology. 2008;**3**(4):201-205

[97] McIver JW, Schulte B, Stein F-U, Matsuyama T, Jotzu G, Meier G, et al. Light-induced anomalous hall effect in graphene. Nature Physics. 2019;**16**:1-4

[98] Erhard N, Seifert P, Prechtel L, Hertenberger S, Karl H, Abstreiter G, et al. Ultrafast photocurrents and THz generation in single InAs-nanowires. Annalen der Physik. 2013;**525**(1–2): 180-188

[99] Kastl C, Karnetzky C, Brenneis A, Langrieger F, Holleitner A. Topological insulators as ultrafast Auston switches in on-chip THz-circuits. IEEE Journal of Selected Topics in Quantum Electronics. 2017;**23**(4):1-5

[100] Kastl C, Karnetzky C, Karl H, Holleitner AW. Ultrafast helicity control of surface currents in topological insulators with near-unity fidelity. Nature Communications. 2015;**6**(1):6617

[101] Seifert P, Vaklinova K, Kern K, Burghard M, Holleitner A. Surface state-dominated photoconduction and THz generation in topological Bi2Te2Se nanowires. Nano Letters. 2017;**17**(2): 973-979

[102] Zimmermann P, Holleitner AW. On-site tuning of the carrier lifetime in silicon for on-chip THz circuits using a focused beam of helium ions. Applied Physics Letters. 2020;**116**(7):073501

[103] Prechtel L, Song L, Schuh D, Ajayan P, Wegscheider W, Holleitner AW. Time resolved ultrafast photocurrents and terahertz generation in freely suspended graphene. Nature Communications. 2012;**3**(1):1-7

[104] Brenneis A, Schade F, Drieschner S, Heimbach F, Karl H, Garrido JA, et al. THz-circuits driven by photo-thermoelectric, gate-tunable graphene-junctions. Scientific Reports. 2016;**6**(1): 35654

[105] Parzinger E, Hetzl M, Wurstbauer U, Holleitner AW. Contact morphology and revisited photocurrent dynamics in monolayer MoS 2. Npj 2D. Materials and Applications. 2017;**1**(1):1-8

[106] Prechtel L, Song L, Manus S, Schuh D, Wegscheider W, Holleitner AW. Time-resolved picosecond photocurrents in contacted carbon nanotubes. Nano Letters. 2011; **11**(1):269-272

[107] Prechtel L, Padilla M, Erhard N, Karl H, Abstreiter G, Anna Fontcuberta I, et al. Time-resolved photoinduced thermoelectric and transport currents in GaAs nanowires. Nano Letters. 2012;**12**(5):2337-2341

[108] Karnetzky C, Zimmermann P, Trummer C, Sierra CD, Wörle M, Kienberger R, et al. Towards femtosecond on-chip electronics based on plasmonic hot electron nano-emitters. Nature Communications. 2018;**9**(1):2471

[109] Th Gerrits HAM, Van Den Berg J, Hohlfeld LB, Rasing T. Ultrafast precessional magnetization reversal by picosecond magnetic field pulse shaping. Nature. 2002;**418**(6897):509-512

[110] Garzon S, Ye L, Webb RA, Crawford TM, Covington M, Kaka S. Coherent control of nanomagnet dynamics via ultrafast spin torque pulses. Physical Review B. 2008;**78**(18): 180401

[111] Wilson RB, Yang Y, Gorchon J, Lambert C-H, Salahuddin S, Bokor J. Electric current induced ultrafast demagnetization. Physical Review B. 2017;**96**(4):045105

[112] Wang Z, Pietz M, Walowski J, Förster A, Lepsa MI, Münzenberg M. Spin dynamics triggered by subterahertz magnetic field pulses. Journal of Applied Physics. 2008; **103**(12):123905

[113] Yang Y, Wilson RB, Gorchon J, Lambert C-H, Salahuddin S, Bokor J. Ultrafast magnetization reversal by picosecond electrical pulses. Science Advances. 2017;**3**(11):e1603117

[114] Jhuria K, Hohlfeld J, Pattabi A, Martin E, Cordova AYA, Shi X, et al. Spin–orbit torque switching of a ferromagnet with picosecond electrical pulses. Nature Electronics. 2020;**3**(11): 680-686

[115] Polley D, Pattabi A, Rastogi A, Jhuria K, Diaz E, Singh H, et al. Picosecond spin-orbit torque induced coherent magnetization switching in a ferromagnet. arXiv preprint arXiv: 2211.08266. 2022

[116] Kato YK, Myers RC, Gossard AC, Awschalom DD. Current-induced spin polarization in strained semiconductors. Physical Review Letters. 2004;**93**(17): 176601

[117] Yoshioka K, Kumada N, Muraki K, Hashisaka M. On-chip coherent frequency-domain thz spectroscopy for electrical transport. Applied Physics Letters. 2020;**117**(16): 161103

[118] Jingbo W, Mayorov AS, Wood CD, Mistry D, Li L, Muchenje W, et al. Excitation, detection, and electrostatic manipulation of terahertz-frequency range plasmons in a two-dimensional electron system. Scientific Reports. 2015; **5**(1):15420

[119] Gallagher P, Yang C-S, Lyu T, Tian F, Kou R, Zhang H, et al. Quantum-critical conductivity of the Dirac fluid in graphene. Science. 2019;**364**(6436): 158-162

[120] Jingbo W, Sydoruk O, Mayorov AS, Wood CD, Mistry D, Li L, et al. Time-domain measurement of terahertz frequency magnetoplasmon resonances in a two-dimensional electron system by the direct injection of picosecond pulsed currents. Applied Physics Letters. 2016; **108**(9):091109

[121] Shaner EA, Lyon SA. Time-resolved impulse response of the magnetoplasmon resonance in a two-dimensional electron gas. Physical Review B. 2002;**66**(4):041402

[122] Russell C, Swithenbank M, Wood CD, Burnett AD, Li L, Linfield EH, et al. Integrated on-chip thz sensors for fluidic systems fabricated using flexible polyimide films. IEEE Transactions on Terahertz Science and Technology. 2016;**6**(4):619-624

[123] Mittleman DM, Hunsche S, Boivin L, Nuss MC. T-ray tomography. Optics Letters. 1997;**22**(12):904-906

Chapter 2

Terahertz Spectroscopy in Advanced Materials Science

Seiji Kojima

Abstract

Materials science is the interdisciplinary field to study material properties and their functionality on the basis of physics, chemistry, metallurgy, and mineralogy. Vibrational spectroscopy such as infrared spectroscopy and Raman spectroscopy is a powerful tool to investigate characteristic atomic vibrations. Especially, in the terahertz frequency range, vibrational modes are related to collective atomic vibrations reflecting interatomic/molecular interactions, characteristic units, and medium range order. Recent progress of terahertz vibrational spectroscopy using terahertz-time-domain spectroscopy, terahertz time-domain spectroscopic ellipsometry, and far-infrared spectroscopy is reviewed in advanced materials science on glassy and crystalline pharmaceuticals, ferroelectrics, and polar metallic materials. Using the terahertz spectra, phonons, polaritons, and conduction electrons of these materials are discussed.

Keywords: terahertz, infrared, spectroscopy, ferroelectrics, pharmaceuticals, phonon, polariton

1. Introduction

In materials science, vibrational spectroscopy is the important and valuable tool to investigate the dynamical properties related to atomic bonds, interactions, and structures. Infrared (IR) and Raman scattering spectroscopies have been extensively used to study various elementary excitations such as phonon, polariton, magnon, exciton, plasmon, boson peak, etc. The most of IR studies reported the frequency-dependent absorbance or transmittance, while the real and imaginary parts of dielectric constants were not reported well. The coherent terahertz generation technique using a femtosecond pulse laser enables the unique determination of a complex dielectric constant, and terahertz time domain spectroscopy has attracted much attention [1, 2].

The vibrational modes of a material are divided into two groups, namely internal and external modes. The internal modes are related to the vibration of a molecule or a structural unit, and its mode frequency is usually more than 6 THz. The external modes are related to the vibration of more than two molecules or lattice vibration with medium or long range order, and its mode frequency is usually less than 6 THz. Therefore, the external modes have been studied by THz spectroscopy. In this chapter, Recent progress of terahertz vibrational spectroscopy using THz-time-domain spectroscopy (THz-TDS) [3], THz time-domain spectroscopic ellipsometry (THz-TDSE) [4], and far-infrared spectroscopy is reviewed in advanced materials science on the following three kinds of materials. (1) Pharmaceuticals: studies of terahertz vibrational spectroscopy on polymorphism of crystalline states and glassy states on typical pharmaceuticals [5]. (2) Ferroelectric materials: studies of terahertz vibrational spectroscopy on optical phonons, ferroelectric soft mode, and phonon-polariton dispersion relation on typical ferroelectric crystals for optical application [6]. (3) Polar metallic materials: studies of terahertz vibrational spectroscopy on the correlation between polar lattice instability and free carrier electrons injected by heterovalent doping to quantum paraelectrics [7].

2. Terahertz spectroscopy in advanced materials science

The structure of a crystal has translational symmetry and the X-ray diffraction patterns of powdered crystals show many sharp peaks. While, glassy, amorphous, and non-crystalline solids have the disordered structure with no translational symmetry and the X-ray diffraction pattern consists of only diffuse peaks by the lack of three dimensional periodicity. When the structure of a crystal has a center of symmetry or centrosymmetric, the mutual exclusion principle holds between Raman and IR selection rules. Raman active modes are IR inactive, and IR active modes are Raman inactive. Some vibrational modes are only Raman active, while other modes are only IR active. Therefore, both infrared and Raman measurements are important. In contrast, the structures of non-centrosymmetric crystals, glassy or non-crystalline materials do not have a center of symmetry, and there is no mutual exclusion principle [2].

2.1 Pharmaceuticals

The most of pharmaceuticals have polymorphic and polyamorphic features originated from multi-basin structure in the energy landscape model. The energy of these basins is nearly degenerate, and the inter-basin transition easily occurs by the change of temperature, pressure, stress, and chemical conditions. The vibrational properties in the THz frequency range are very sensitive to the difference between polymorphic and polyamorphic states. Therefore, in the study of polymorphic and polyamorphic natures in pharmaceuticals, the measurements of both THz-TDS and Raman scattering spectroscopy are important. These two spectroscopic methods have the complimentary selection rules which give new insights into the detailed information on dynamical and static properties related to crystalline and noncrystalline structures, medium range order, and interatomic/intermolecular interactions [5].

THz transmission spectra were measured by the conventional transmission THz-TDS system (Tochigi Nikon, RT-10,000), where the low temperature grown GaAs photoconductive antennas were used for both the emitter and detector in the frequency range from 0.2 to 4.0 THz. The Raman scattering spectra were measured in the frequency range from 0.2 to 36 THz using a double-grating spectrometer with the photon-counting system (Horiba-JY, U-1000). The imaginary part of Raman susceptibility, $\chi''(\nu)$, is extracted by the equation,

$$\chi''(\nu) = \frac{I(\nu)}{I_0\{n(\nu)+1\}}, n(\nu) = \frac{1}{exp\left(\frac{h\nu}{k_B T}\right) - 1}, \tag{1}$$

Here $I(\nu)$ and $n(\nu)$ are the scattering intensity and the Bose-Einstein factor, respectively. I_0 is a constant which depends on the experimental condition.

2.1.1 Vibrational properties of glassy and crystalline pharmaceuticals

Indapamide ($C_{16}H_{16}ClN_3O_3S$, IND) is 4-chloro-N-(2-methyl-2,3-dihydroindol-1-yl)-3-sulfamoylbenzamide. On the treatment of hypertension, the commercially available type IND exerts spasmolytic effects on blood vessels and reduces the blood pressure. Its glass-forming tendency is relatively strong. IND undergoes a liquid-glass transition at a glass transition temperature, T_g = 376 K, which is relatively higher than room temperature [8]. The unit cell of a crystalline phase of commercial type IND contains four IND molecules (Z = 4). The crystal framework contains two types of cavities. The cavity encapsulates water molecules in non-stoichiometric. Since the water molecules are weakly bounded in these cavities, they easily come out and in from the IND crystalline framework [9].

The thin plates of crystalline and glassy samples prepared by the melt-quench method were measured by THz-TDS between 0.2 and 3.0 THz at room temperature [10]. **Figure 1a** and **b** show the real and imaginary parts of the dielectric constants of the commercial type glassy and crystalline states of IND. In the observed frequency range, several sharp phonon peaks were observed in a crystalline sample, these modes can be attributed to inter- or intra-molecular vibrational modes in a crystalline state. However, no sharp phonon peak was observed in a glassy state and the broad response reflects the distribution of bond lengths and bond angles in the random structure of a glassy state.

For the comparison of the THz-TDS spectra with the Raman scattering spectra, the imaginary parts of Raman susceptibility, $\chi"(\nu)$, of crystalline and glassy states are shown in **Figure 2**. As similar with the THz-TDS spectra, a crystalline sample shows several sharp phonon peaks, while a glassy sample shows only broad response and no sharp phonon peak. In a crystalline state, the discrepancy of the phonon peak frequencies was clearly observed between $\varepsilon"(\nu)$ observed by THz-TDS in **Figure 1b** and $\chi"(\nu)$ observed by Raman scattering in **Figure 2**. The origin of the discrepancy is the mutual exclusion principle between Raman and IR activities [11]. Consequently, it is concluded that the point symmetry of the crystalline IND has a center of symmetry. This result is consistent with the monoclinic crystal structure with the point group *2/m* determined by the X-ray diffraction experiment [12]. The existence of a center of

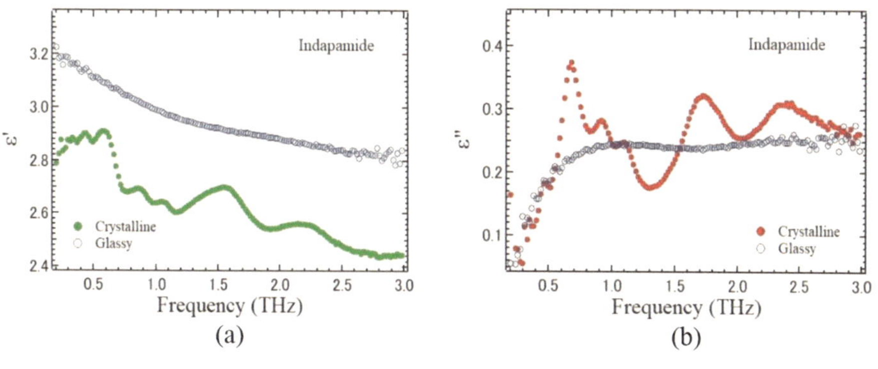

Figure 1.
(a) Real and (b) imaginary parts of dielectric constant of crystalline and glassy indapamide.

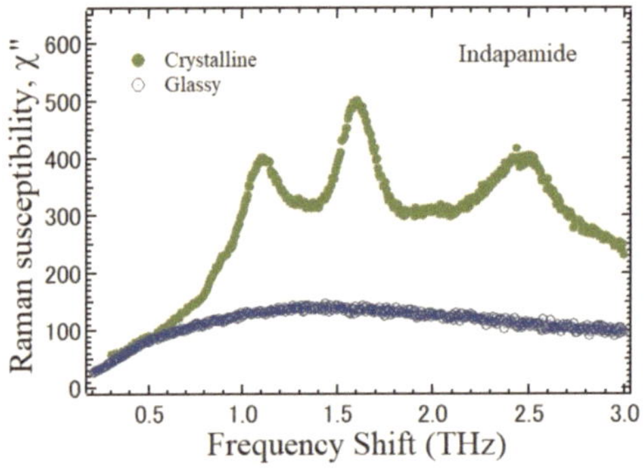

Figure 2.
Imaginary parts of Raman susceptibility of crystalline and glassy indapamide.

symmetry was also examined by THz-TDS and Raman spectroscopy in crystalline γ-form indometachine [11] and crystalline RS-ketoprofen [13].

In contrast, ε"(ν) and χ"(ν) spectra show very broad response and no sharp peak below 3.0 THz in a glassy state. If the frequency range is extended to the IR region, many internal modes can be observed. In a glassy glycerol, the temperature dependence of the intermolecular hydrogen bond length was estimated from the frequency of O-H stretching mode at about 105 THz [14].

2.1.2 Low-energy excitation in a glassy state

As the common nature of the low-energy dynamics of glassy, amorphous, and noncrystalline materials, a broad and asymmetric peak at few THz has been observed by Raman scattering, far-infrared spectroscopy, neutron and X-ray inelastic scattering [15, 16]. It is called a boson peak and a thermal boson peak has been also observed as a broad peak in the temperature dependence of C_p/T^3 curve at a few Kelvin, where C_p is the heat capacity [17]. Both boson peaks are related to $g(\nu)/\nu^2$, where g(ν) is the vibrational density of states (VDoS), g(ν). The strong correlation between the boson peak energy and the shear modulus has been reported [18]. The molecular dynamical simulation reported that a boson peak is originated from the Ioffe-Regal limit of transverse acoustic mode and its peak energy has a linear relation with the shear modulus [19].

In the far-IR spectroscopy, a boson peak is observed by,

$$\frac{\kappa(\nu)}{\nu} \propto C_{IR}(\nu)\frac{g(\nu)}{\nu^2}. \tag{2}$$

Here, κ(ν) is the real part of a refractive index, and C_{IR} (ν) is the IR-vibration coupling constant [20]. The boson peak was fitted by the following log-normal function under the assumption that the IR-vibration coupling constant does not depend on frequency.

$$\frac{k(\nu)}{\nu} = I_0 \exp\left\{-\frac{[ln\,(\nu/\nu_{BP})]^2}{2\sigma^2}\right\}. \tag{3}$$

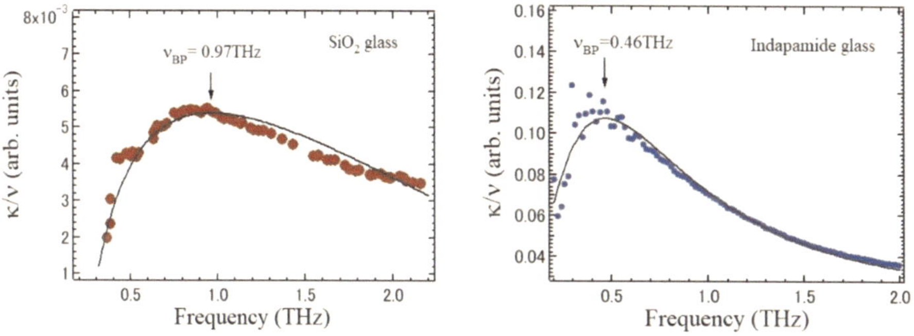

Figure 3.
Boson peaks observed by THz-TDS. (a) SiO_2 glass (ν_{BP} = 0.9 THz), (b) glassy indapamide (ν_{BP} = 0.4 THz) at room temperature. Solid lines show fitted values by Eq. (2). Solid circles denote observed values.

Here I_0 is a constant, ν_{BP} and σ are the boson peak frequency and boson peak width, respectively. **Figure 3a** and **b** show boson peaks at room temperature observed by THz-TDS on SiO_2 glass ($\nu_{BP} \approx 0.9$ THz) [21] and glassy indapamide ($\nu_{BP} \approx 0.4$ THz) [10], respectively. The observation of boson peaks by THz-TDS was reported in various glasses such as PMMA [22], indomethacin [11], glucose [23], and lysozyme [24]. The study on boson peaks of glassy materials will give new insights into the understanding of disordered materials with no translational symmetry.

2.2 Quantum paraelectrics

The ABO_3-type perovskite oxides have been extensively used for potential applications such as capacitor, ferroelectric random access memory, piezoelectric actuator, ultrasonic transducer, pyroelectric sensor, and surface acoustic wave filter. $BaMO_3$ (M = Ti, Zr) compounds are very important in pure and applied sciences [25]. Barium titanate, $BaTiO_3$ has been widely used as capacitor. $BaTiO_3$ undergoes a ferroelectric phase transition at 403 K from a prototypic cubic phase with the point group $m\bar{3}m$ to the tetragonal phase with *4 mm,* where $\bar{3}$ and m are three-fold rotation-inversion axis and mirror plane, respectively. Its physical properties have been extensively studied.

Barium zirconate, $BaZrO_3$ has attracted much attention by a high-temperature proton conductor [26] and a dielectric material for use in wireless communications applications [27]. $BaZrO_3$ does not undergo any phase transition down to 2 K [28]. Its dielectric constant increases on cooling, while a ferroelectric instability is suppressed by quantum fluctuations at very low temperatures. It is called quantum paraelectricity. As quantum paraelectrics, $SrTiO_3$, $KTaO_3$, and $CaTiO_3$ are known. The cubic perovskite phase has a center of symmetry with the point group $m\bar{3}m$, and infrared-active ferroelectrtic soft modes are Raman inactive. Although the low-frequency infrared-active modes play a dominant role in ferroelectric instability, the experimental study in the far-infrared spectroscopy has been not enough.

The real and imaginary parts of a dielectric constant of a $BaZrO_3$ crystal was measured between 8 and 300 K by the transmission of THz-TDS [29, 30]. In **Figure 4a**, the peak of the imaginary part at 1.9 THz was observed at 8 K. The TO mode frequency was determined by the fitting the imaginary part of a dielectric constant $\varepsilon(\nu)$ using the following damped harmonic oscillator (DHO) model,

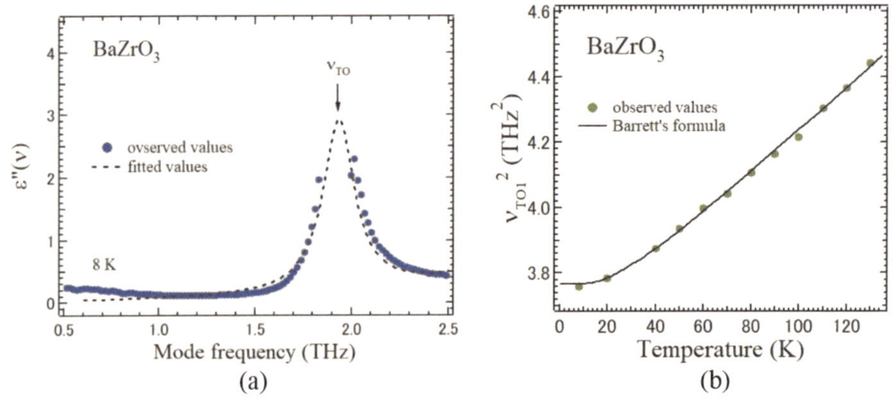

Figure 4.
(a) Imaginary part of dielectric constant of a $BaZrO_3$ crystal at 8 K. (b) Temperature dependence of the square of the TO1 mode frequency. The solid line shows the fitted curve by the Barrett's formula [29, 30].

$$\varepsilon(\nu) = \varepsilon(\infty) + \sum_j \frac{\Delta\varepsilon_j \nu_j^2}{\nu_j^2 - \nu^2 - i\nu\Gamma_j}, \tag{4}$$

where $\varepsilon(\infty), \Delta\varepsilon_j, \nu_j$, and Γ_j are the high-frequency limit of a dielectric constant, oscillator strength, frequency and damping factor of j th oscillator mode, respectively.

The temperature dependence of the squared mode frequency of the lowest TO1 mode is shown in **Figure 4b**. The squared soft mode frequency ν_{TO}^2 of a quantum paraelectric obeys the following Barrett's formula [31].

$$\nu_{TO}^2 = A\left[T_s coth\left(\frac{T_S}{T}\right) - T_{01}\right]. \tag{5}$$

Here, A, T_S, and T_{01} are constant, saturation temperature, and classical Curie temperature, respectively. The lowest frequency TO1 infrared-active phonon mode exhibits a significant softening, while the quantum effects lead to saturation of the soft mode frequency.

3. Phonon-polariton dispersion relation

The coupled excitation between photons and other elementary excitations is called polariton [32, 33]. The remarkable frequency vs. wavevector dispersion relation of polariton has attracted much attention in basic science to clarify the vibrational dynamics of condensed matters such as phonons, excitons, and plasmons. Polar phonon modes, which are infrared active, couple to photons at frequencies and the mixed excitation is called phonon-polariton. Phonon polaritons have been used for optical applications such as a tunable Raman laser, tunable terahertz (THz) radiation source. The dispersion relation of phonon-polariton has been extensively studied by infrared spectroscopy, spontaneous, hyper Raman and stimulated Raman scattering spectroscopy [34]. However, Raman inactive phonon-polariton cannot be measured by spontaneous and stimulated Raman scattering spectroscopy. The phonon-polariton of Raman inactive phonons can be measured only by Raman spectroscopy and hyper-Raman scattering spectroscopy. Therefore, the study of Raman inactive polaritons has been insufficient.

According to Huang's analysis [31], the dispersion relation of phonon polariton was defined by the equation,

$$\varepsilon\left(\vec{k},\nu\right) = \frac{c^2k^2}{(2\pi\nu)^2}, \tag{6}$$

where ν, $\vec{k}$, c, and $\varepsilon(\vec{k},\nu)$ are the polariton frequency, polariton wavenumber, light velocity, and dielectric constant of a medium, respectively. **Figure 5** shows the dispersion relation of phonon polariton on the existence of two infrared active optical modes with the mode frequencies, ν_{TO1} = 6.0 THz, ν_{LO1} = 6.9 THz, ν_{TO2} = 19.8 THz, and ν_{LO2} = 24.9 THz. As the polariton wavenumber k tends to zero, the lower polariton branch tends to $\nu = ck/2\pi\sqrt{\varepsilon(0)}$, where $\varepsilon(0)$ is the lowest frequency limit of a dielectric constant, while the middle and upper branches tend to LO1 mode frequency, ν_{LO1}, and LO2 mode frequency, ν_{LO2}, respectively. When the polariton wavenumber k tends to infinity, the lower and middle branches tend to TO1 mode frequency, ν_{TO1}, and TO2 mode frequency $\nu = \nu_{TO2}$, respectively, while the upper branch tends to $\nu = ck/2\pi\sqrt{\varepsilon(\infty)}$, where $\varepsilon(\infty)$ is the highest frequency limit of a dielectric constant.

In the observation of phonon-polariton by Raman scattering, the conservation law of wavevectors holds,

$$\pm\vec{k}(\nu) = \vec{k}_i(\nu_i) - \vec{k}_s(\nu_s), \tag{7}$$

where $\vec{k}$, $\vec{k}_i$, and $\vec{k}_s$ are the wavevectors of polariton, incident light, and scattered light, respectively. ν, ν_i, and ν_s are the frequencies of polariton, incident light, and

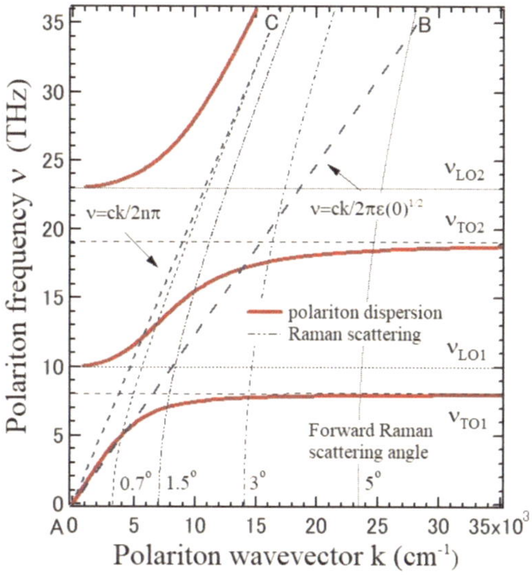

Figure 5.
Dispersion relation of phonon-polariton with two optical modes. The lines AB and AC denote $\nu = ck/2\pi\sqrt{\varepsilon(0)}$ *and* $\nu = ck/2\pi\sqrt{\varepsilon(\infty)}$*, respectively. The dotted lines show the observable region of the forward Raman scattering with a fixed scattering angle.*

scattered light, respectively. The polariton wavenumber k is given by the wavenumbers of the incident light k_i, scattered light k_s, and the scattering angle θ between them.

$$k^2 = k_i^{\,2} + k_s^{\,2} - 2k_i k_s \cos\theta. \tag{8}$$

The dispersion relation of polariton is also determined by the forward Raman scattering at very small scattering angles as shown in **Figure 5**. While, it is impossible to measure the dispersion relation in the region, $\nu > ck/2\pi\sqrt{\varepsilon(0)}$ by Raman scattering [35]. In contrast, the measurement by infrared spectroscopy has no such restriction and covers all area of ν vs.$\vec{k}$ space.

3.1 Raman active phonon-polariton

3.1.1 Bismuth titanate

On the ferroelectric memories, bismuth layer-structure materials have currently become important in the research and development of ferroelectric devices. These compounds are also very important as the most suitable samples to study the mechanism in ferroelectric layer-structures. Bismuth titanate, $Bi_4Ti_3O_{12}$ undergoes a ferroelectric phase transition from a paraelectric tetragonal phase with the point group *4/mmm* to a ferroelectric monoclinic phase with the point group *m* at T_C = 948 K, where *4* and *m* are four-fold rotation axis and mirror plane, respectively [36]. The ferroelectric soft optic mode was observed by the Raman scattering. On heating from room temperature, the lowest frequency mode at 0.84 THz with A'(*x*,*z*) symmetry softens markedly and disappears above T_C [37].

In the determination of the dispersion relation of phonon-polariton by the transmission measurement of THz-TDS, the wavenumber $k(\nu)$ of phonon-polariton is calculated using the phase delay φ as a function of the polariton frequency ν by the following equation,

$$\varphi = \{k(\nu) - 2\pi\nu/c\}d, k(\nu) = 2\pi\,\nu\,n(\nu)/c, \tag{9}$$

where c, d, and $n(\nu)$ are the light velocity, a thickness of a sample, and the real part of a refractive index of a sample, respectively.

The anisotropy of polariton dispersion relations between A'(*x*,*z*) and A"(*y*) symmetry modes was also successfully determined using the *c* plate simply by switching the polarization direction of an incident beam from $E\|a$ to $E\|b$, respectively [38]. The observed polariton dispersion relations are consistently reproduced by the calculation using the following Kurosawa's formula (**Figure 6**).

$$\varepsilon(k,\nu) = \frac{c^2k^2}{(2\pi\nu)^2} = \varepsilon(\infty)\prod_{j=1}^{N}\frac{\nu_{LOj}^2 - \nu^2}{\nu_{TOj}^2 - \nu^2} \tag{10}$$

Since the measured frequency is below 3.0 THz, we used the following approximation to fit the data:

$$\frac{c^2k^2}{(2\pi\nu)^2} = \varepsilon(1)\prod_{j=1}^{2}\frac{\nu_{LOj}^2 - \nu^2}{\nu_{TOj}^2 - \nu^2}. \tag{11}$$

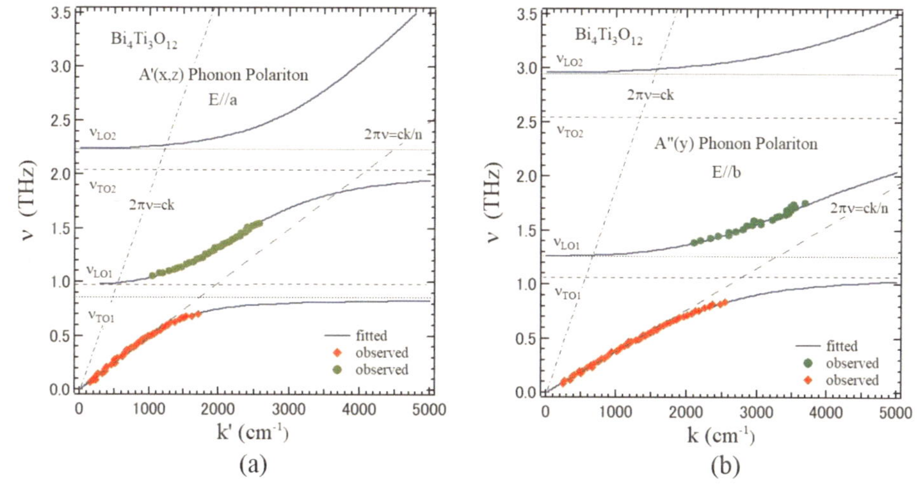

Figure 6.
Dispersion relations of (a) A'(x,z) symmetry and (b) A'(y) symmetry phonon-polaritons of a $Bi_4Ti_3O_{12}$ crystal.

The values of these parameters are listed in **Tables 1** and **2**. The observed values of polariton frequency were well fitted by Eq. (11), because the anharmonicity is negligible in a $Bi_4Ti_3O_{12}$ crystal at room temperature.

3.1.2 Lithium niobate, $LiNbO_3$

Lithium niobate, $LiNbO_3$, has been used for many kinds of devices by their excellent piezoelectric, pyroelectric, and nonlinear optical properties. $LiNbO_3$ is a colorless uniaxial crystal and undergoes a ferroelectric phase transition at a high Curie temperature, T_C = 1210°C from paraelectric ($R\bar{3}c$ with three-fold rotation-inversion

Mode	**A'(*x*,*z*), *E*‖*a***	**A"(*y*), *E*‖*b***
TO1	0.84 THz	1.08 THz
LO1	0.98 THz	1.26 THz
TO2	2.04 THz	2.55 THz
LO2	2.24 THz	2.96 THz

Table 1.
Fitting values of mode frequency in Eq. (11).

	A'(*x*,*z*), *E*‖*a*	**A"(*y*), *E*‖*b***
$\varepsilon(0)$	79.2	149.0
$\varepsilon(1)$	49.0	81.0
$\varepsilon(\infty)$	6.76	6.76

Table 2.
Fitting values of dielectric constant in Eq. (11).

axis, $\bar{3}$) to ferroelectric (*R3c* with three-fold rotation axis, *3*) phases [39]. Lattice dynamics have been extensively studied by Raman scattering [40], infrared spectroscopy, and theoretical calculations. In a rhombohedral ferroelectric phase, the symmetry of the optical modes at the Γ point of Brillouin zone is given by.

$$4A_1(z) + 9E(x, y) + 5A_2. \tag{12}$$

Here, $A_1(z)$ and $E(x,y)$ modes are Infrared and Raman active, while A_2 modes are silent modes, which are infrared and Raman inactive. The symmetry of a ferroelectric soft mode is $A_1(z)$. According to the Raman scattering study, the temperature dependence of the lowest frequency $A_1(z)$ mode became overdamped far below T_C. This fact indicates that the anharmonicity of a ferroelectric soft mode is very high even at room temperature. The Raman scattering spectrum of $A_1(z)$ modes observed at the $a(cc)\bar{a}$ backward scattering is shown in **Figure 7** [41]. Therefore, the damping of $A_1(z)$ modes is not negligible.

For the analysis of such anharmonic modes, it is necessary to discuss the damping of phonon using the imaginary part of a dielectric constant. When a dielectric constant, $\varepsilon(\vec{k}, \nu)$, is a complex number, then a wavenumber of polariton, *k* or a frequency, *ν*, can be also a complex number. Since the infrared spectroscopy considers spatial decay, *ν* is defined as a real number, while *k* is defined as a complex number [32]. The complex wavenumber of polariton, $k(\nu)$, is defined by the equation,

$$k(\nu) = k'(\nu) + ik''(\nu). \tag{13}$$

Considering the following relation between the complex dielectric constant and the complex refractive index,

$$\varepsilon(k, \nu) = \{n(k, \nu) + i\kappa(k, \nu)\}^2. \tag{14}$$

The polariton dispersion relation for the real and imaginary parts of the complex *k* (*ν*) was derived from Eqs. (13) and (14),

$$k'(\nu) = 2\pi/c\, n(k, \nu), k''(\nu) = 2\pi/c\, \kappa(k, \nu) \tag{15}$$

The polariton dispersion of anharmonic $A_1(z)$ modes of a congruent $LiNbO_3$ crystal was measured by FIRSP [41] as shown in **Figure 8**. The polariton dispersion relation was determined by the complex refractive constant measured in the condition *E*//*c*.

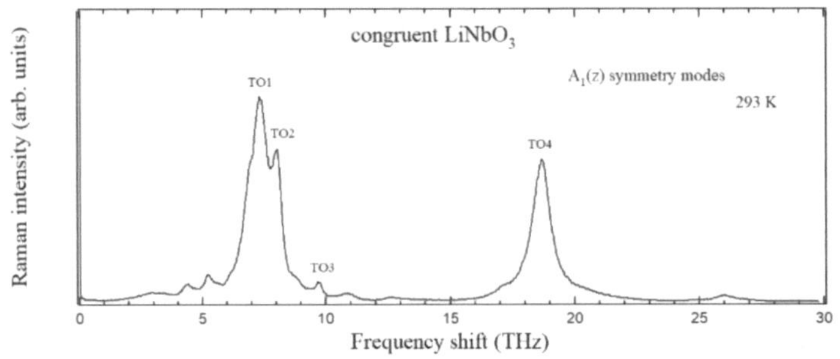

Figure 7.
Raman scattering spectrum of $A_1(z)$ modes observed at the $a(cc)\bar{a}$ backward scattering.

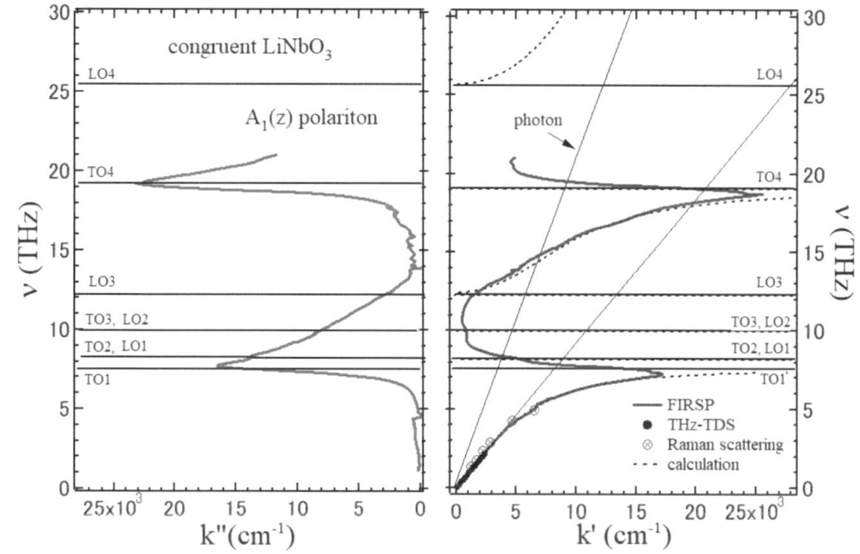

Figure 8.
Dispersion relations of phonon-polariton with real and imaginary parts of wavenumber of a $LiNbO_3$ crystal are shown in right and left hands sites, respectively. TOj and LOj (j = 1,2,3) denote the three TO and three LO modes of the T_{1u} symmetry, respectively [42]. Closed and open circles denote the observed values by THz-TDS [43] and Raman scattering [44, 45], respectively.

Mode	Far-IR ellipsometry	Raman scattering	First principles calculation [45]
A_1(TO1)	7.54 THz	7.63 THz	7.18 THz
A_1(TO2)	8.17 THz	8.35 THz	8.67 THz
A_1(TO3)	9.97 THz	10.0 THz	10.6THz
A_1(TO4)	19.0 THz	19.0 THz	18.3THz

Table 3.
A_1(TO) mode frequency.

Figure 8 shows the real and imaginary parts of polariton wavenumber versus the polariton frequency with $A_1(z)$ symmetry. The result by forward Raman scattering measurements [43, 44] was also plotted for the comparison. The dotted lines in **Figure 8** denote the calculated dispersion relation with no phonon damping. Closed open circles in **Figure 8** denote the values observed by the forward Raman scattering measurement. These observed values of Raman scattering and FIRSP are in good agreement with those determined by the first principles calculations [45] within experimental uncertainty as shown in **Table 3**.

3.2 Raman inactive phonon-polariton

The tolerance factor of perovskite $SrTiO_3$ (STO) is 1.0. The STO is known as the typical quantum paraelectric [46] as same as $BaZrO_3$ in Section 2.2. The space group symmetry is a cubic $Pm\bar{3}m$ with the center of symmetry at room temperature. The optical vibrational modes at the Γ point of the Brillouin zone are $3T_{1u} + T_{2u}$. The T_{2u} modes are silent modes, which are infrared inactive and Raman inactive. The $3T_{1u}$

modes are Raman inactive while infrared active and hyper-Raman active. The T_{1u} modes were studied by the far infrared spectroscopy [47], THz-TDS [48, 49], and hyper-Raman scattering [50, 51].

For the study of the polariton dispersion relation of the three T_{1u} modes in the large frequency range up to 36 THz, the infrared reflectivity spectrum of a [001] STO plate was measured in the range from 0.9 and 36 THz as shown in **Figure 9** [52]. The typical polariton dispersion curve was observed in the vicinity of the lowest TO1 mode at ν_{TO1} = 2.6 THz. the results of hyper-Raman scattering [50, 51] were also plotted in **Figure 9** for the comparison. Inoue et al. measured only the highest-frequency polariton dispersion higher than the highest-frequency LO3 mode at ν_{LO3} = 23.7 THz [51]. While, Denisov et al. measured both the lowest-frequency dispersion curve lower than the lowest TO1 mode frequency and the highest-frequency dispersion curve higher than the highest LO3 mode [50]. The polariton dispersion measured by the hyper-Raman scattering measurements [50, 51] is in a good agreement within the experimental uncertainty with the results the infrared reflection measurement [52] in the frequency range below 2.6 THz and above 23.7 THz.

In hyper-Raman scattering spectroscopy, the conservation law holds among the wave vectors of an incident, $\vec{k}_i$, scattered light, $\vec{k}_s$, and polariton, $\vec{k}$,

$$\pm\vec{k}(\nu) = 2\vec{k}_i(\nu_i) - \vec{k}_s(\nu_s), \tag{16}$$

where ν_i and ν_s are frequencies of incident and scattered light from a sample, respectively. The frequencies of $\nu = 2\nu_i - \nu_s$ and ν_s is approximately equal to a double of ν_i. According to the Eq. (16), the dispersion relation is observable only in the limited region due to the birefringence between the refractive indices of fundamental of an incident light and second harmonic wavelengths of scattered light. Especially, the

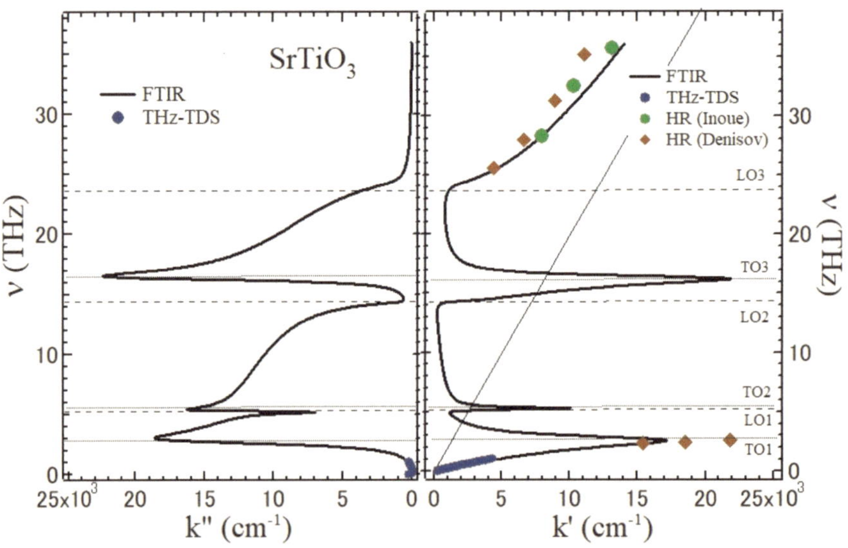

Figure 9.
Dispersion relations of phonon-polariton with real and imaginary parts of wavenumber of a $SrTiO_3$ crystal are shown in right and left hands sites, respectively. TOj and LOj (j = 1, 2, 3) denote the three TO and three LO modes of the T_{1u} symmetry, respectively [52]. The results of FTIR [52] and THz TDS [49] are shown by the solid line and the closed circles, respectively. The values of the hyper-Raman scattering [50, 51] are shown by the closed circles and closed diamonds.

observation of the lowest frequency polariton of a soft mode with a small $\vec{k}_s$ is impossible by this birefringence. Consequently, the hyper-Raman scattering of phonon-polariton is not suitable to study the ferroelectric soft mode.

In Section 3, the polariton dispersion of Raman active modes in $Bi_4Ti_3O_{12}$, $LiNbO_3$, and Raman inactive modes of $SrTiO_3$ are described. Other ferroelectrics such as $LiTaO_3$, $BaZrO_3$, and β-$Gd_2(MoO_4)_3$ were also studied by THz-TDS [53].

4. Polar metallic materials: ferroelectric instability and conduction electrons

The possibility of ferroelectric metals was suggested by Anderson and Blount in 1965 that V_3Si and other metallic transitions may be "ferroelectric" by the appearance of a polar axis [54]. It was written that while free electrons screen out the electric field completely, they do not interact very strongly with the transverse optical phonons and the Lorentz local fields lead to ferroelectricity, since umklapp processes are forbidden as $k \to 0$. Recently the relation between ferroelectricity/polar distortion and metallic conductivity has attracted much attention in various kinds of metallic materials [55]. The A or B-site doped $SrTiO_3$ crystals has been extensively studied by the significant changes of physical properties related to metallic conduction by injected free electrons and lattice distortions [56, 57].

For the measurements of a materials with a high dielectric constant, the high reflectivity causes the difficulty in a transmission THz measurement. On reflection measurements using a conventional THz-TDS system, the determination of the phase delay of the reflected beam from a sample is difficult, because the reference beam from the surface of a sample cannot be accurately measured. Therefore, for the measurement of a complex high dielectric constant, the terahertz time-domain spectroscopic ellipsometry (THz-TDSE) using the reflection from a sample to be studied has been developed. In this section, the ferroelectric soft mode and conduction electrons studied by THz-TDSE is described [58, 59].

4.1 Ferroelectric instability

The complex dielectric constants of $SrTiO_3$ crystals with heterovalent B-site doping by Nb^{5+} ions were investigated to clarify the correlation between the ferroelectric soft optic mode and the injected carrier electrons. By the Fourier transformation of the time dependence of electric fields of p- and s-polarization components of reflected THz waves, the real and imaginary parts of a dielectric constant at room temperature were determined as shown in **Figure 10a** and **b**, respectively [59]. The dielectric response in the THz range includes a ferroelectric soft optic mode and dynamics of carrier electrons. The former is dominant near the ferroelectric soft mode frequency around 3 THz, which is well reproduced by the damped harmonic oscillator (DHO) model. While, in the frequency region lower than 1.5 THz, the dynamics of carrier electrons are dominant. Usually, the dielectric response of carrier electrons has been discussed using the Drude model. However, the accurate THz range spectra needs the modification of this model [60]. The Drude-Smith model assuming a single collision [61] has been used to analyze THz spectra. The complex dielectric constant of the Drude-Smith model is given by,

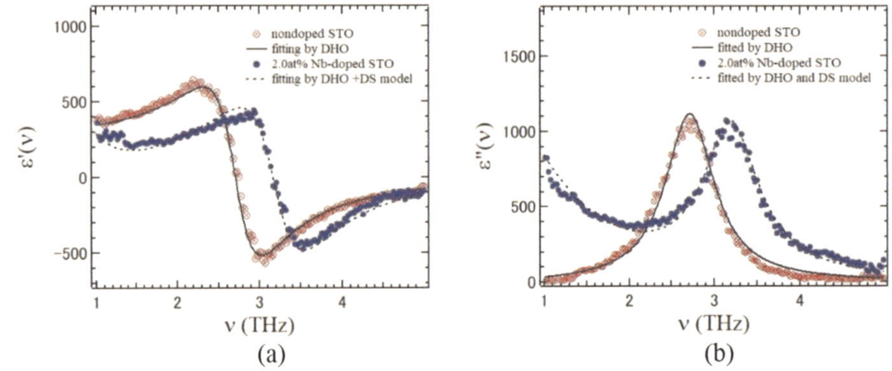

Figure 10.
(a) Real and (b) imaginary parts of the complex dielectric constant of Nb doped $SrTiO_3$ crystals measured by THz-TDSE at room temperature.

$$\varepsilon_{DS}(\nu) = \varepsilon_b + \frac{i2\pi\nu_p^2\tau}{\nu(1 - i2\pi\nu\tau)}\left(1 + \frac{c}{1 - i2\pi\nu\tau}\right), \quad -1 \le c \le 0, \tag{17}$$

where ε_b and ν_p, are the background dielectric constant and plasma frequency, respectively. The case of $c = 0$ is the Drude model, and as the constant c decreases from zero, the back scattering of electrons increases.

The observed real and imaginary parts of complex dielectric constants of Nb-doped $SrTiO_3$ were fitted by the summation of the DHO mode responsible for a soft polar mode and the Drude-Smith model responsible for free carriers as shown by the dotted lines in **Figure 10a** and **b**. Since in an undoped $SrTiO_3$ crystal has no free carriers, the spectra were fitted only by the DHO model as shown in the solid lines in **Figure 10a** and **b**.

4.2 Correlation between soft mode and conduction electrons

The lowest frequency TO1 mode is a ferroelecric soft optic mode in $SrTiO_3$. The soft mode frequency and plasma frequency are plotted as a function of the carrier concentration n_c in Nb doped $SrTiO_3$ crystals as shown in **Figure 11a**. It is found that both soft mode frequency and plasma frequency increase as the carrier concentration increases.

In polar semiconductors, it is known that longitudinal optical phonons couples to plasmons. At the appropriate carrier density, the free-carrier plasma frequency is comparable to the phonon frequency, and the coupling between plasmons and phonons occurs [62]. Then the two new normal modes are strong admixtures of phonons and plasmons. The macroscopic point symmetry $m\bar{3}m$ of undoped $SrTiO_3$ crystals has a center of symmetry, while there is the possibility of local symmetry breaking [63] in doped $SrTiO_3$ crystals. For Nb doped $SrTiO_3$ the calculated plasma frequency was plotted as a function of the root of carrier concentration dependence in the relatively low concentration region as shown in **Figure 11b**. However, the concentration of **Figure 11a** is much higher than that of **Figure 11b**, and the plasma frequency is more than ten times of a soft mode frequency. Therefore, the effect of such a coupling is negligible in the present study.

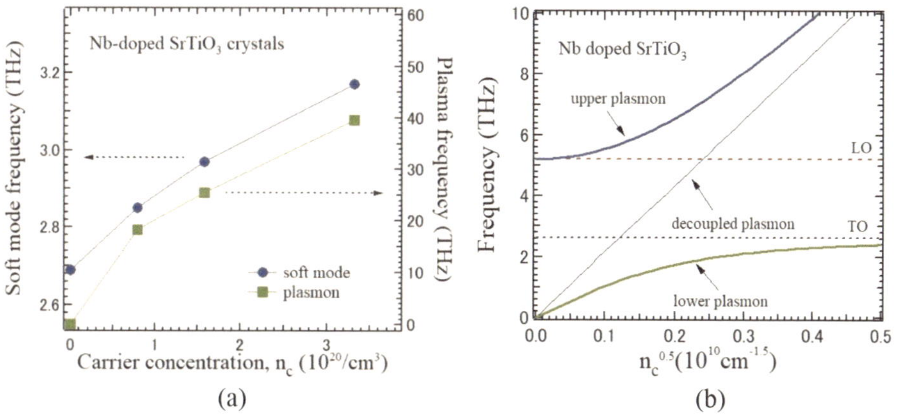

Figure 11.
(a) The soft mode frequency and plasma frequency are plotted as a function of the carrier concentration n_c *in Nb doped* $SrTiO_3$ *crystals. (b) The coupled plasma frequency against the root of* n_c.

According to the Cochran's theory on ferroelectric instability [64], the square of the soft mode frequency, ν_S, is proportional to the difference between short range (SR) and long range (LR) interactions.

$$\nu_S^2 \propto [\text{SR interaction}] - [\text{LR interaction}] \tag{18}$$

The LR electrostatic fields favoring the polar structure are expected to be screened by the free carriers. The charge carriers essentially suppress the LR interaction of the attractive Coulomb force and increase a ferroelectric soft mode frequency. **Figure 12** shows the relation between the square of a soft mode frequency and the free carrier concentration in doped $SrTiO_3$ crystals. It is found that the square of a ferroelectric soft mode frequency linearly increases as the carrier concentration increases. It is concluded that the carrier electron concentration suppresses the ferroelectirc instability and enhances the increase of a soft mode frequency.

However, the suppression of polar instability by free carriers markedly depends on materials reflecting the strength of electron-phonon coupling. The first-principles

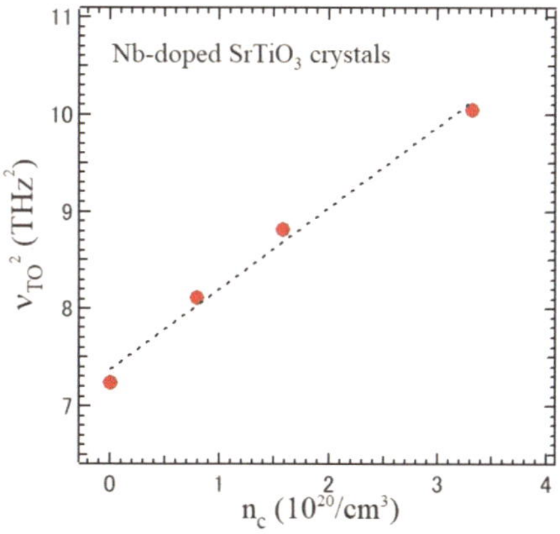

Figure 12.
Square of the ferroelectric soft mode frequency, ν_{TO}, *as a function of carrier concentration,* n_c, *of Nb doped* $SrTiO_3$ *crystals. The dotted line shows the linear relation.*

calculations showed that substantial polar displacements in n-doped $LiNbO_3$ are much more stable than those in n-doped $BaTiO_3$. This fact indicates that the electron-phonon coupling of $BaTiO_3$ is much stronger than that of $LiNbO_3$ [65]. Recently, it was reported that nonadiabatic Born effective charges (naBECs) can be defined in metallic compounds [66]. For doped $SrTiO_3$, the calculation of naBECs of the sublattices in $SrTiO_3$ suggested significant changes in the mode frequencies of polar phonon. The detailed theoretical study may clarify the interaction of the carrier electrons with ferroelectric instability.

5. Summary

Advanced terahertz vibrational spectroscopy using THz-time-domain spectroscopy (THz-TDS), THz time-domain spectroscopic ellipsometry (THz-TDSE), and far-infrared spectroscopy is reviewed. In pharmaceuticals, it is suggested that both THz-TDS and Raman scattering spectroscopy on polymorphic and polyamorphic states are important to clarify their structure and dynamics. As to the phonon-polariton dispersion relation, the detection of anisotropy and the dispersion relation for real and imaginary parts of polariton wavenumber are discussed in technologically important ferroelectric materials. The comparison between THz time-domain spectroscopy and forward Raman scattering is discussed. Recently polar metallic materials with metallic conduction and polar distortion have attracted much attention. In the heterovalent doped quantum paraelectric strontium titanate, the metallic conduction and polar instability coexist. Both dynamical features are studied by THz time-domain ellipsometry. The metallic polar instability is analyzed by the damped harmonic oscillator model for a ferroelectric soft mode and the Drude-Smith model for free carriers.

Acknowledgements

Author is thankful for the collaboration and discussions for M.W. Takeda, M. Maczka, H. Kitahara, S. Nishizawa, T. Iwamoto, Y. Satou, T. Hoshina, T. Mori, H. Igawa, T. Shibata, S. Koda, Y. Kobayashi.

Author details

Seiji Kojima
Division of Materials Science, University of Tsukuba, Tsukuba, Ibaraki, Japan

*Address all correspondence to: kojima@ims.tsukuba.ac.jp

IntechOpen

© 2023 The Author(s). Licensee IntechOpen. This chapter is distributed under the terms of the Creative Commons Attribution License (http://creativecommons.org/licenses/by/3.0), which permits unrestricted use, distribution, and reproduction in any medium, provided the original work is properly cited.

References

[1] Neu J, Schmuttenmaer CA. Tutorial: An introduction to terahertz time domain spectroscopy (THz-TDS). Journal of Applied Physics. 2018; **124**(23):231101. DOI: 10.1063/1.5007683

[2] Kojima S, Shibata T, Igawa H, Mori T. Broadband terahertz time-domain spectroscopy: Crystalline and glassy drug materials. IOP Conference Series: Material Science Engineering. 2014; **54**(1):012001. DOI: 10.1088/1757-899X/54/1/012001

[3] van Exter M, Grischkowsky D. Optical and electronic properties of doped silicon from 0.1 to 2 THz. Applied Physics Letters. 1990;**56**(17):1694-1696. DOI: 10.1063/1.103120

[4] Ino Y, Shimano R, Svirko Y, Kuwata-Gonokami M. Terahertz time domain magneto-optical ellipsometry in reflection geometry. Physical Review B. 2004;**70**(15):155101. DOI: 10.1103/PhysRevB.70.155101

[5] Kojima S, Mori T, Shibata T, Kobayashi Y. Broadband terahertz time-domain and low-frequency Raman spectroscopy of crystalline and glassy pharmaceuticals. Pharmaceutical Analytical Acta. 2015;**6**(8):401. DOI: 10.4172/21532435.1000401

[6] Kojima S. Broadband terahertz spectroscopy of phonon-polariton dispersion in ferroelectrics. Photonics. 2018;**5**(4):55. DOI: 10.3390/photonics5040055

[7] Kojima S. Terahertz time-domain spectroscopic ellipsometry of metallic strontium titanate. Ferroelectrics. 2022; **601**(1):1-8. DOI: 10.1080/00150193.2022.2130762

[8] Wojnarowska Z, Grzybowska K, Hawelik L, Dulski M, Wrzalik R, et al. Molecular dynamics, physical stability and solubility advantage from amorphous Indapamide drug. Molecular Pharmaceutics. 2013;**10**:3612-3627. DOI: 10.1021/mp400116q

[9] Ghugare P, Dongre V, Karmuse P, Rana R, Singh D, et al. Solid state investigation and characterization of the polymorphic and pseudopolymorphic forms of indapamide. Journal of Pharmaccutical Biomedical Analysis. 2010;**51**(3):532-540. DOI: 10.1016/j.jpba.2009.09.003

[10] Kobayashi Y, Shibata T, Mori T, Kojima S. Terahertz time-domain spectroscopy and low-frequency Raman scattering of crystalline and glassy pharmaceutical Indapamide. International Letter Chemical and Physical Astrology. 2015;**46**:16-22. DOI: 10.56431/p-khvpdt

[11] Shibata T, Mori T, Kojima S. Low-frequency vibrational properties of crystalline and glassy indomethacin probed by terahertz time-domain spectroscopy and low-frequency Raman scattering. Spectrochimica Acta Part A: Molecular and Biomolecular Spectroscopy. 2015;**150**:207-211. DOI: 10.1016/j.saa.2015.05.059

[12] Smrkolj M, Meden A. Crystal structure of indapamide determined from powered diffraction data. Pharmazie. 2006;**61**(12):999-1004

[13] Shibata T, Igawa H, Kim TH, Mori T, Kojima S. Glass transition dynamics of anti-inflammatory ketoprofen studied by Raman scattering and terahertz time-domain spectroscopy. Journal of Molecular

Structure. 2014;**1062**:185-188. DOI: 10.1016/j.molstruc.2014.01.050

[14] Kojima S. Anomalous behavior of the O—H stretch vibrational mode in the liquid-glass transition of glycerol. Journal of Molecular Structure. 1993; **294**(1):193-196. DOI: DOI. 10.1016/0022-2860(93)80347-X

[15] Nakayama T. Boson peak and terahertz frequency dynamics of vitreous silica. Reports on Progress in Physics. 2002;**65**:1195-1242. DOI: 10.1088/0034-4885/65/8/203

[16] Kojima S. Low energy excitations in borate glass. In: Wongchoosuk C, editor. Characteristics and Applications of Boron [Internet]. London: IntechOpen; 2022. Available from: https://www.intechopen.com/chapters/83194. DOI: 10.5772/intechopen.106650

[17] Zeller RC, Pohl RO. Thermal conductivity and specific heat of non-crystalline solids. Physical Review B. 1971;**4**:2029-2040. DOI: 10.1103/PhysRevB.4.2029

[18] Shintani H, Tanaka H. Universal link between the boson peak and transverse phonons in glass. Nature Materials. 2008;**7**:870-877. DOI: 10.1038/nmat2293

[19] Ioffe AF, Regel AR. Non-crytalline, amorphous, and liquid electronic semiconductors. Progress in. Semiconductors. 1960;**4**:237-291

[20] Galeener FL, Sen PN. Theory for the first-order vibrational spectra of disordered solids. Physical Review B. 1978;**17**(4):1928-1933. DOI: 10.1103/PhysRevB.17.1928

[21] Kojima S, Kitahara H, Nishizawa S, Yang YS, Takeda MW. Terahertz time-domain spectroscopy of low-energy excitations in glasses. Journal of Molecular Structure. 2005;**744–747**: 243-246. DOI: 10.1016/j.molstruc.2004.10.045

[22] Kojima S, Nishizawa S, Takeda MW. Complex dielectric constants of boson peaks of PMMA probed by terahertz time domain spectroscopy. Journal of Molecular Structure. 2003;**651–653**:285-288. DOI: 10.1016/S0022-2860(02)00641-5

[23] Kabeya M, Mori T, Fujii Y, Koreeda A, Lee BW, Ko JH, et al. Boson peak dynamics of glassy glucose studied by integrated terahertz-band spectroscopy. Physical Review B. 2016; **94**(22):224204. DOI: 10.1103/PhysRevB.94.224204

[24] Mori T et al. Detection of boson peak and fractal dynamics of disordered systems using terahertz spectroscopy. Physical Review E. 2020;**102**(2):022502. DOI: 10.1103/PhysRevE.102.022502

[25] Helal MA, Kojima S. Brillouin scattering and first-principles studies of $BaMO_3$ (M = Ti, Zr, and Cu) perovskites. Materials. 2022;**15**:6747. DOI: 10.3390/ma1519674

[26] Ostapchuk T, Petzelt J, Zelezny V, Paskin A, Pokorny J, Drbohlav I, et al. Origin of soft-mode stiffening and reduced dielectric response in $SrTiO_3$ thin films. Physical Review B. 2002; **66**(23):235406. DOI: 10.1103/PhysRevB.66.235406

[27] Kreuer KD. Proton-conducting oxides. Annual Review of Materials Research. 2003;**33**:333. DOI: 10.1146/annurev.matsci.33.022802.091825

[28] Akbarzadeh AR, Kornev I, Malibert C, Bellaiche L, Kiat JM. Combined theoretical and experimental study of the low-temperature properties of $BaZrO_3$. Physical Review B. 2005;

72(20):205104. DOI: 10.1103/PhysRevB.72.205104

[29] Helal MA, Mori T, Kojima S. Softening of infrared-active mode of perovskite $BaZrO_3$ proved by terahertz time-domain spectroscopy. Applied Physics Letters. 2015;**106**(18):182904. DOI: 10.1063/1.4919913

[30] Helal MA, Mori T, Kojima S. Terahertz time-domain spectroscopy and Raman scattering studies of incipient ferroelectric $BaZrO_3$. Ferroelectrics. 2016;**499**(1):107-114. DOI: 10.1080/00150193.2016.1173502

[31] Barrett JH. Dielectric constant in perovskite type crystals. Physical Review. 1952;**86**(1):118-120. DOI: 10.1103/PhysRev.86.118

[32] Born M, Huang K. Dynamical Theory of Crystal Lattices. Oxford: Oxford University Press; 1954

[33] Brandmuller J, Claus R, Merten L. Light Scattering by Phonon-polaritons. Berlin Heidelberg: Springer-Verlag; 1975. DOI: 10.1007/BFb0048910

[34] Kojima S, Maczka M. Broadband phonon-polariton dispersion relation of ferroelectric $LiTaO_3$ crystals. Ferroelectrics. 2018;**533**:124-131. DOI: 10.1080/00150193.2018.1470825

[35] Kojima S, Nakamura T. Observations of low-frequency phonon-polariton in barium sodium niobate. Japanese Journal of Applied Physics. 1980;**19**:L609-L611

[36] Subbarao EC. Ferroelectricity in $Bi_4Ti_3O_{12}$ and its solid solutions. Physical Review. 1961;**122**(3):804-807. DOI: 10.1103/PhysRev.122.804

[37] Kojima S, Shimada S. Soft mode spectroscopy of bismuth titanate single crystals. Physica B. 1996;**219–220**:617-619. DOI: 10.1016/0921-4526(95)00830-6

[38] Kojima S, Tsumura T, Takeda MW, Nishizawa S. Far-infrared phonon-polariton dispersion probed by terahertz time domain spectroscopy. Physical Review B. 2003;**67**(3):035102. DOI: 10.1103/PhysRevB.67.035102

[39] Matthias BT, Remeika JP. Ferroelectricity in the ilmenite structure. Physical Review. 1949;**76**(12):1886-1887. DOI: 10.1103/PhysRev.76.1886.2

[40] Surovtsev NV, Pugachev AM, Malinovsky VK, Shebanin AP, Kojima S. Low-frequency Raman spectra in $LiNbO_3$: Within and beyond the standard paradigm of ferroelectric dynamics. Physical Review B. 2005;**72**(10):104303. DOI: 10.1103/PhysRevB.72.104303

[41] Kojima S, Kanehara K, Hoshina T, Tsurumi T. Optical phonons and polariton dispersions of congruent $LiNbO_3$ studied by far-infrared spectroscopic ellipsometry and Raman scattering. Japanese Journal of Applied Physics. 2016;**55**:10TC02

[42] Kojima S, Kitahara H, Wada Takeda M, Nishizawa S. Terahertz time domain spectroscopy of phonon-polaritons in ferroelectric lithium niobate crystals. Japanese Journal of Applied Physics. 2002;**41**(11S):7033-7037. DOI: 10.1143/JJAP.41.7033

[43] Kojima S, Nakamura T. Low frequency phonon polaritons in several ferroelectrics. Ferroelectrics. 1981;**37**:677-680. DOI: 10.1080/00150198108223519

[44] Kojima S. Raman scattering study of composition variation in the low-frequency $A_1(z)$ modes of $LiNbO_3$.

Journal of the Korean Physical Society. 1998;**32**(9):552-555

[45] Sanna S, Neufeld S, Rüsing M, Berth G, Zrenner, Schmidt WG. Raman scattering efficiency in $LiTaO_3$ and $LiNbO_3$ crystals. Physical Review B. 2015;**91**(22):224302. DOI: 10.1103/PhysRevB.91.224302

[46] Muller KA, Burkard H. $SrTiO_3$: An intrinsic quantum paraelectric below 4 K. Physical Review B. 1979;**19**(7):3593-3602. DOI: 10.1103/PhysRevB.19.3593

[47] Spitzer WG, Miller RC, Kleinman DA, Howarth LE. Far infrared dielectric dispersion in $BaTiO_3$, $SrTiO_3$, and TiO_2. Physical Review. 1962;**126**(5):1710-1721. DOI: 10.1103/PhysRev.126.1710

[48] Matsumoto N, Fujii T, Kageyama K, Takagi H, Nagashima T, Hangyo M. Measurement of the soft-mode dispersion in $SrTiO_3$ by terahertz time-domain spectroscopic ellipsometry. Japanese Journal of Applied Physics. 2002;**48**(9S1):09KC11. DOI: 10.1143/JJAP.48.09KC11

[49] Kojima S, Mori T. Terahertz time-domain spectroscopy of Raman inactive phonon-polariton in strontium titanate. Ferroelectrics. 2016;**499**:100-106. DOI: 10.1080/00150193.2016.1173500

[50] Denisov VN, Mavrin BN, Podobedev VB, Sterin KE. Hyper Raman scattering by polaritons in centrally symmetric $SrTiO_3$ crystal. JETP Letters. 1980;**31**(2):102-105

[51] Inoue K, Asai N, Samejima T. Observation of the phonon polariton in the centrosymmetric crystal of $SrTiO_3$ by Hyper Raman scattering. Journal of the Physical Society of Japan. 1980;**48**(5):1787-1788. DOI: 10.1143/JPSJ.48.1787

[52] Kojima S, Maczka M. Raman inactive phonon-polariton dispersion in quantum paraelectric $SrTiO_3$ crystals studied by FTIR. Ferroelectrics. 2018;**524**(1):1-8. DOI: 10.1080/00150193.2018.1431493

[53] Kojima S, Mori T. Terahertz time-domain spectroscopy of infrared active soft mode and phonon-polariton dispersion. Ferroelectrics. 2016;**500**(1):183-202. DOI: 10.1080/00150193.2016.1214522

[54] Anderson PW, Blount EI. Symmetry consideration on martensitic transformations: "ferroelectric" metals. Physical Review Letters. 1965;**14**(7):21. DOI: 10.1103/PhysRevLett.14.217

[55] Zhou WX, Ariando A. Review on ferroelectric/polar metals. Japanese Journal of Applied Physics. 2020;**59**(5):SI0802. DOI: 10.35848/1347-4065/ab8bbf

[56] Bednorz JG, Müller KA. $Sr_{1-x}Ca_xTiO_3$: An XY quantum ferroelectric with transition to randomness. Physical Review Letters. 1984;**52**(25):2289-2292. DOI: 10.1103/PhysRevLett.52.2289

[57] Rischau CW et al. A ferroelectric quantum phase transition inside the superconducting dome of $Sr_{1-x}Ca_xTiO_{3-\delta}$. Nature Physics. 2017;**13**:64. DOI: 10.1038/nphys4085

[58] Kojima S, Iwamoto T, Satou Y. Study of A-site substituted quantum paraelectric strontium titanate crystals by terahertz time-domain spectroscopic ellipsometry. Japanese Journal of Applied Physics. 2020;**59**:SPPA02. DOI: 10.35848/1347-4065/aba2c1

[59] Kojima S, Iwamoto T, Satou Y. Lattice instability of B-site substituted $SrTiO_3$ crystals studied by terahertz time-domain spectroscopic ellipsometry. Japanese Journal of Applied Physics.

2021;**59**:SPPA02. DOI: 10.35848/1347-4065/ac13da

[60] Lourembam J, Srivastava A, Laovorakiat C, Rotella H, Venkatesan T, Chia EEM. New insights into the diverse electronic phases of a novel vanadium dioxide polymorph: A Terahertz Spectroscopy Study. Scientific Reports. 2015;**5**:9182. DOI: 10.1038/srep09182

[61] Smith NV. Classical generalization of the Drude formula for the optical conductivity. Physical Review B. 2001; **64**(15):155106. DOI: 10.1103/PhysRevB.64.155106

[62] Mooradian A, McWhorter AL. Polarization and intensity of Raman scattering from plasmons and phonons in gallium arsenide. Physical Review Letters. 1967;**19**(15):849-852. DOI: 10.1103/PhysRevLett.19.849

[63] Pugachev AM, Kovalevskii VI, Surovtsev NV, Kojima S, Prosandeev SA, Raevski IP, et al. Broken local symmetry in paraelectric $BaTiO_3$ proved by second harmonic generation. Physical Review Letters. 2012;**108**(24):247601. DOI: 10.1103/PhysRevLett.108.247601

[64] Cochran W. Crystal stability and the theory of ferroelectricity. Advances in Physics. 1960;**9**(36):387. DOI: 10.1080/00018736000101229

[65] Xia C, Chen Y, Chen H. Coexistence of polar displacements and conduction in doped ferroelectrics: An ab initio comparative study. Physical Review Materials. 2019;**3**(5):054405. DOI: 10.1103/PhysRevMaterials.3.054405

[66] Dreyer CE, Coh S, Stengel M. Nonadiabatic Born effective charges in metals and the Drude weight. Physical Review Letters. 2022;**128**(9):095901. DOI: 10.1103/PhysRevLett.128.095901

Chapter 3

Systematic Characterization of High-Dielectric Constant Glass Materials Using THz-TDS Technique

Osamu Wada, Doddoji Ramachari, Chan-Shan Yang, Takashi Uchino and Ci-Ling Pan

Abstract

High-dielectric constant glasses are prerequisite for developing terahertz (THz) components and systems. Oxyfluorosilicate (OFS) glasses have been developed and their THz properties have been characterized by using THz-time domain spectroscopy (THz-TDS) measurements. High-dielectric constant (8–13) and low loss (6–9/cm) properties in the THz region have been demonstrated and their dielectric properties have been studied using the single oscillator-based model through a comparison with other multi-component silicate oxide glasses. Unified single oscillator model, which can distinguish the electronic and ionic contributions to the dielectric property, has been applied in this analysis. The physical origin of the dielectric constant enhancement and the importance of interplay between the electronic polarizability and ionicity in high-dielectric constant glasses have been revealed. This study has demonstrated the usefulness of THz-TDS technique for characterizing dielectric properties of multi-component glasses in detail.

Keywords: terahertz time domain spectroscopy (THz-TDS), dielectric constant, multi-component glasses, electronic/ionic polarizability, polarization ionicity

1. Introduction

Since its early demonstrations in 1980–1990's, terahertz (THz) time-domain spectroscopy (THz-TDS) [1–5] has been employed to study a wide variety of crystalline and non-crystalline materials, e.g., dielectrics [6], semiconductors [7, 8], polymers [9], biological molecules [9], liquid crystals [10], aqueous solutions [11], and many others. The THz-TDS technique has a variety of advantages [9, 11]. For example, a compact system can be built by adopting compact, convenient-to-use components such as photoconductive antenna (PCA) devices as THz sources as well as detectors that can be excited by a sub-picosecond mode-locked laser. Both reflection and transmission configurations can be used depending on the sample nature. Original THz signal obtained is in time domain, but it can be Fourier transformed into frequency domain over a wide bandwidth. The amplitude and phase information are independently acquired and precise material characteristics can be obtained. Thus, THz-TDS has become a powerful tool for detailed physical/chemical characterization and spectroscopic chemical assessment of materials.

The characteristic vibrational properties of solid materials such as glasses are often found in the THz frequency range (usually 0.1–10 THz). However, glass properties studied so far are limited to some silicate oxide and chalcogenide glasses [12–16]. Some glasses have shown relatively high transparency in the THz frequency range [14–16], and those are potentially useful for the application to a variety of photonic devices, components, and systems in the THz frequency band.

A number of multi-component silicate glasses [17] and various chalcogenide glasses [18] have been developed for optical frequency applications owing to their high chemical/mechanical stability and desirable optical properties such as high refractive index, high transparency, and high optical nonlinearity in visible and mid-infrared wavelength regions. Toward THz applications, further development in glass materials with high-refractive index and low loss is desired. One possibility is oxyfluoride (OF) glasses [19–21], in which non-radiative transitions are suppressed with corresponding high-efficiency luminescence and transparency. Recently, oxyfluorosilicate (OFS) glasses containing metal fluorides have been investigated as they can exhibit combined properties of oxides and fluorides such as a low melting temperature, high chemical/mechanical resistance, low phonon energy, and attractive optical characteristics including high refractive index [19–21]. Multi-component design to increase the electronic and ionic polarizabilities can be useful for realizing high refractive index in the THz frequency range. This also suggests a possibility of high optical nonlinearity in the THz frequency range [22], which is attractive for developing active devices such as switching and wavelength conversion devices. Thus, OFS glasses incorporating appropriate metal oxides/fluorides are expected to provide essential glass materials for THz applications. The THz dielectric properties and their physical nature of such OFS glasses, however, have not been systematically investigated so far.

In the following, we describe the characterization of high-dielectric constant and low loss properties of novel OFS glasses, which are revealed by the application of the THz-TDS technique. First, we describe the THz dielectric properties of two series of multi-component glasses: ZNbKLSNd and PbNKLSNd OFS glasses [23]. Then, we discuss the physical mechanism of the dielectric constant enhancement through a comparative study involving several other multi-component silicate oxide and chalcogenide glasses by using the conventional, simplified single oscillator model [24] and also by developing and applying the more precise, unified single oscillator model [25]. The results show how the dielectric properties of glasses are controlled by the electronic polarizability and ionicity of the glasses, which also demonstrates the usefulness of THz-TDS technique in the glass material characterization.

2. Experimental methods

2.1 THz characterization and analysis techniques

Figure 1 shows the schematic diagram of the transmission-type photoconductive-antenna-based THz-TDS setup employed in this work. Details can be found in our

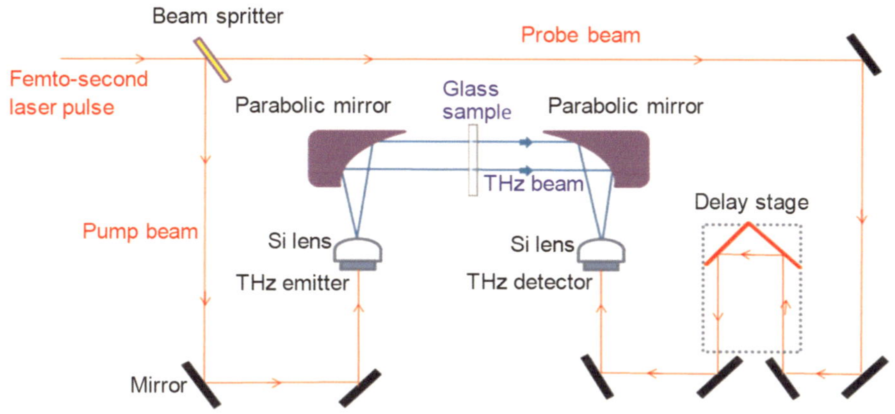

Figure 1.
Schematic diagram of transmission-type photoconductive-antenna-based THz-TDS setup used in this study.

previous work [7, 26]. Briefly, the ultrafast laser used was a Ti:sapphire femto-second laser (Spectra Physics, Tsunami) that routinely generated ~60 fs pulses with an average power of 380 mW at 82 MHz. A dipole-type photoconductive antenna (PCA) fabricated on low-temperature molecular-beam-epitaxy-grown GaAs (LT-GaAs) was biased by a periodic step-function-type signals and was excited by the ultrafast laser pulses to generate THz pulses. The generated THz wave was collected using a high-resistivity Si lens and was collimated by gold-coated parabolic mirror into a highly directional beam. After passing through the glass sample, the THz pulse beam was focused onto an identical LT-GaAs PCA as the detector using an identical combination of parabolic mirror and Si lens, as shown in **Figure 1**. The electric field of the transmitted THz pulse induces a voltage across the detector antenna and is mapped by photoconductively shorting the gap. When the detector was gated by a delayed probe laser pulse, a current proportional to the instantaneous field strength of the incident THz pulse can thus be measured using the lock-in technique. The time delay between the pump and probe laser pulse is mechanically scanned by moving a retroreflector with a computer-controlled motor stage. The amplitude and phase of the THz wave were obtained by changing the time delay using this setup. The THz-TDS system was enclosed in a N_2-purged box with a relative humidity of ~4% to minimize THz power absorption due to residual water vapor in the THz beam path. The glass samples with diameter of ~10 mm and thickness of ~1.6 mm were polished on both sides. The samples were attached to a sample holder that has an optical aperture ~8 mm in diameter. An identical clear aperture was used as a reference.

Typical characteristics of the emitted THz wave in this THz-TDS system are shown in **Figure 2**: (a) displays the measured electric field intensity in time domain and (b) shows the frequency spectrum of THz power, which has been retrieved using the fast Fourier transformation (FFT). The effect of nitrogen purge is obvious for minimizing the water absorption. As shown in the spectrum, the dynamic range of the THz-TDS system is ~45 dB at 0.5 THz, and the S/N ratio is maintained in the range of 100–1000 across the band from 0.2 to 1.6 THz.

Figure 3a shows the electric field temporal traces of the transmitted THz wave with and without the glass sample, and **Figure 3b** shows the frequency dependence of the THz phase shift estimated with respect to the reference phase angle (without glass sample). Since the phase shift occasionally showed irregular behavior at high frequency (>0.8 THz, refer to **Figure 3b**) and correspondingly worse S/N ratio, we have

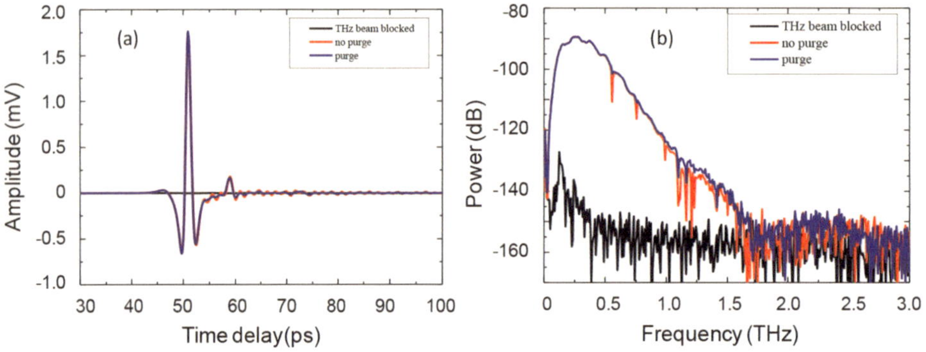

Figure 2.
Measured data of (a) time-domain trace of electric field intensity of emitted THz wave and (b) frequency spectrum of emitted THz power as obtained by FFT.

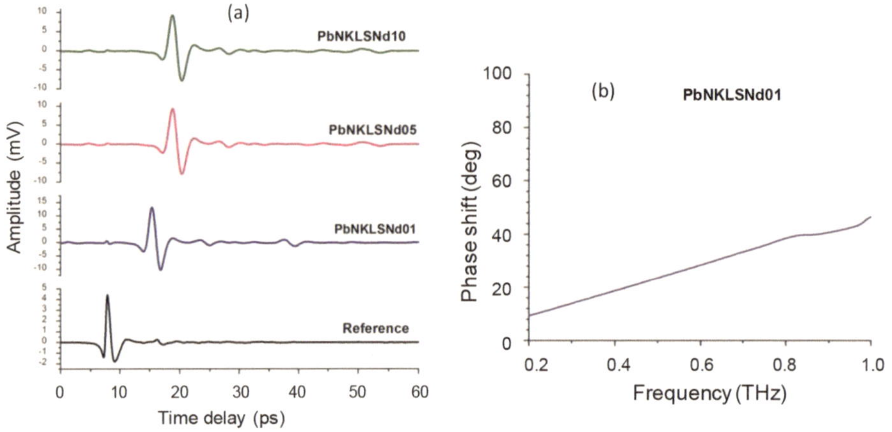

Figure 3.
Measured transmission signals for a representative glass sample (PbNKLSNd glasses). (a) Time domain electric field intensity signals of THz waves transmitted through glass samples and air (reference). (b) THz frequency dependence of phase shift caused by transmission through glass sample.

restricted our measurement frequency range between 0.2 and 0.8 THz for the accuracy.

The THz absorption coefficient $\alpha(\nu)$ and refractive index $n(\nu)$ of the glass samples at the frequency ν are determined by using the following equations [12, 26–28];

$$\alpha(\nu) = \frac{2}{d}\ln\left(\left|\frac{E_{ref}(\nu)}{E_{sam(\nu)}}\right|\frac{[n(\nu)+1]^2}{4n(\nu)}\right) \tag{1}$$

$$n(\nu) = 1 + \frac{c}{2\pi\nu d}\left(\varphi_{sam}(\nu) - \varphi_{ref}(\nu)\right) \tag{2}$$

where $E(\nu)$ and $\phi(\nu)$ are the amplitude and phase of the THz field transmitted through the glass sample (suffix: sam) and the reference medium (air in this case, suffix: ref); c is the speed of light, and d is the thickness of glass sample. The second term inside the natural logarithm in the right hand of Eq. (1) is a Fresnel reflection correction factor, for which the effect of extinction coefficient is neglected because of its insignificant effects in the present measurement frequency range (0.2–0.8 THz).

Refractive indices in the transparent range of optical frequencies were evaluated from the normal incidence reflectivity measurement using Fourier transform infrared (FTIR) spectroscopy (Bruker VERTEX v70) by applying the following simplified equation [24, 29]:

$$n_{opt} = \left(1 + r^{1/2}\right) / \left(1 - r^{1/2}\right), \tag{3}$$

where r is the optical reflectivity. The effect of internal multiple reflection within the sample bulk [29] may influence the reflectivity value in the highly transparent wavelength region, but this effect was confirmed to be negligible in our case, supporting the use of simplified equation as shown above [30].

2.2 Glass sample preparation and data acquisition

The OFS glasses were prepared by the melt-quenching technique [19–21, 23] with two series of chemical compositions; one is ZNbKLSNd glass: $(20 - x)ZnF_2 + 20Nb_2O_5 + 20K_2CO_3 + 10LiF + 30SiO_2 + xNd_2O_3$, and the other is PbNKLSNd glass: $(20\text{-}x)PbF_2 + 5Na_2O + 20K_2CO_3 + 10LiF + 45SiO_2 + xNd_2O_3$, where x = 1, 5, and 10 mol%. The compositions were designed by referring to our prior works on OFS glasses including ZnF_2 [19] and PbF_2 [21]. Different x values were adopted to check the relative effects of the fluorine and modifier (Nd) contents. Batch composition for each glass with ~15 g weight was thoroughly crushed in an agate mortar and its homogeneous mixture was transferred into a platinum crucible and heated at 1250°C for 3 hours in an electric furnace under ambient atmosphere. The melt was then poured onto a preheated brass mold and annealed at 400°C for 10 hrs to remove thermal strains. The glasses were slowly cooled down to room temperature and polished and formed into disks with the thickness of ~1.6 mm for the THz and optical measurements. In order to discuss the dielectric properties and their physical mechanisms, data of several of other multi-component glasses, in particular, several commercial silicate oxide glasses have been collected from the literature. Compositions and basic physical parameters including the molecular weight M, density ρ, and molecular volume V_m of glasses used are summarized in **Table 1**. For OFS glasses, the value of ρ was measured by the Archimedes method. The formula weights of glasses were evaluated using nominal atomic compositions.

Glass	Composition		M (g/mol)	ρ (g/cm^3)	V_m (cm^3/mol)	Ref
ZNbKLSNdx	$(20\text{-}x)ZnF_2 + 20Nb_2O_5 + 20K_2CO_3 + 10LiF + 30SiO_2 + xNd_2O_3$	x = 1	124	3.65	34.0	[23]
		x = 5	133	3.66	36.3	
		x = 10	145	3.54	41.0	
PbNKLSNdx	$(20\text{-}x)PbF_2 + 5Na_2O + 20K_2CO_3 + 10LiF + 45SiO_2 + xNd_2O_3$	x = 1	110	3.72	29.7	[23]
		x = 5	114	3.78	30.2	
		x = 10	119	3.62	32.8	
Silica	SiO_2		60	2.20	27.3	[12, 23]
Pyrex	$80.6SiO_2 + 12.6B_2O_3 + 4.2Na_2O + 2.2Al_2O_3 + 0.04Fe_2O_3 + 0.1CaO + 0.05MgO + 0.1Cl$		62	2.23	27.8	[12, 23]

Glass	Composition	M (g/mol)	ρ (g/cm³)	V_m (cm³/mol)	Ref
BK7	$68.9SiO_2 + 10.1B_2O_3 + 8.8Na_2O + 8.4K_2O + 2.8BaO + 1.0As_2O_3$	65	2.51	26	[12, 23]
SK10	$30.6SiO_2 + 11.7B_2O_3 + 5.0Al_2O_3 + 0.1Na_2O + 48.2BaO + 2.0ZnO + 0.7PbO + 0.8Sb_2O_3 + 0.9As_2O_3$	112	3.64	30.8	[12, 23]
SF10	$35.3SiO_2 + 2.0Na_2O + 2.5K_2O + 55.7PbO + 4.0TiO_2 + 0.5As_2O_3$	153	4.28	35.8	[12, 23]
SF6	$27.7SiO_2 + 0.5Na_2O + 1.0K_2O + 70.5PbO + 0.3As_2O_3$	177	5.18	34.2	[12, 23]
S1	$La_{20}Ga_{20}S_{60}$	61	4.27	14.3	[16]
S2	$La_{16}Ga_{24}S_{60}$	58	4.48	13.0	[16]
S3	$La_{12}Ga_{28}S_{60}$	56	4.11	13.6	[16]
S4	$La_{12}Ga_{28}S_{48}Se_{12}$	61	3.99	15.3	[16]
S5	$La_{12}Ga_{28}S_{39}Se_{21}$	65	4.21	15.4	[16]
S6	$Ge_{33}As_{12}Se_{55}$	76	4.66	16.3	[16]
S7	$Ge_{28}Sb_{12}Se_{60}$	82	4.41	18.6	[16]

Table 1.
Compositions and basic properties of multi-component glasses (reproduced from ref. [23] with the permission of AIP publishing).

3. Results and discussion

3.1 THz refractive index and absorption coefficient dependences on frequency

Figure 4 shows the absorption coefficients, $\alpha(\nu)$, and the refractive indices, $n(\nu)$ of the OFS glasses as a function of the frequency in the frequency range from 0.2 to 0.8 THz. As shown in **Figure 4**, the refractive index increases gradually and the absorption coefficient increases a little more steeply with the increase of frequency in both OFS glasses. General behaviors of frequency dependent refractive index and absorption coefficient as noticed in our results are consistent with prior reports on

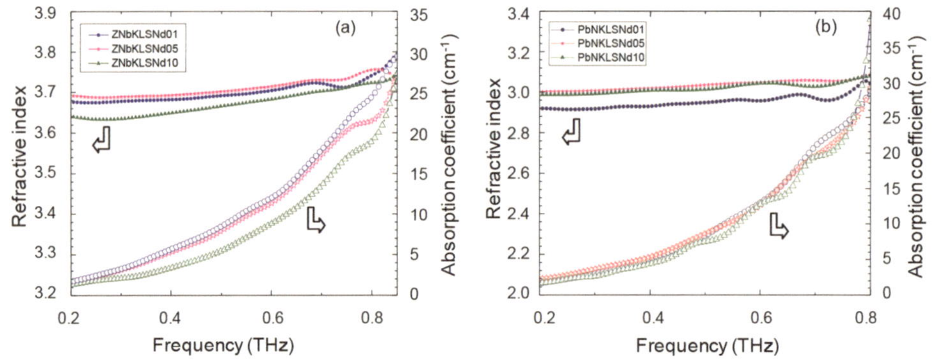

Figure 4.
Refractive index and absorption coeficicent dependences on THz frequency for (a) ZNbKLSNd and (b) PbNKLSNd glasses with different compositions. (reproduced from ref. [23] with the permission of AIP Publishing).

oxide and chalcogenide glasses in the sub-THz frequency band [12–16]. Although weak oscillatory behaviors are observed at higher frequency regions in some samples possibly due to residual reflection/scattering related effects in the system, this is not too severe to affect the evaluations of basic dielectric parameters. The refractive indices measured in those OFS glasses are higher than in most of the silicate glasses. In particular, ZNbKLSNd glasses exhibit refractive indices as high as 3.70 at 0.5 THz, which is the highest among various silicate glasses so far reported and is close to those of La^{3+}: chalcogenide glasses as reported in Ref. [16]. Another important finding in this result is that the OFS glasses exhibit absorption coefficients significantly lower than those of other glasses known as high-refractive index glasses [12].

A comprehensive physical model of far infrared absorption in solids was discussed for various amorphous materials by Strom et al. [31, 32], in which disorder-induced charge fluctuations in the material structure cause the coupling of THz radiation into atomic vibration modes. This model has been applied recently to interpret the THz absorption characteristics in silicate and chalcogenide glasses [12, 16]. In this model, the product of $n(\nu)$ and $\alpha(\nu)$ is shown to follow a power-law relationship as expressed by [13, 14, 16]:

$$n(\nu)\alpha(\nu) = K(h\nu)^{\beta}, \quad (4)$$

where K and β are material-dependent parameters of each glass and h is the Planck's constant. We have analyzed the present OFS glasses on the basis of this relationship. **Figure 5** shows the product $n\alpha$ plotted as a function of the THz frequency for ZNbKLSNd and PbNKLSNd glasses. The $n\alpha$ versus frequency behaviors are in fairly good agreement with the slope of 2 as indicated by dashed lines drawn in the figures. The absorption parameter K has therefore been determined for $\beta = 2$ in our analysis on the present OFS glasses. This is consistent with previous data reported for a number of amorphous materials [31, 32], including silicate [12] and chalcogenide [16] glasses, which supports the disorder-induced-charge fluctuation model for interpreting the sub-THz absorption characteristics in a wide variety of glasses. The value of K is associated with the magnitude of incoherent (not maintaining the local charge neutrality) atomic-charge fluctuation induced by structural disorders [13]. The

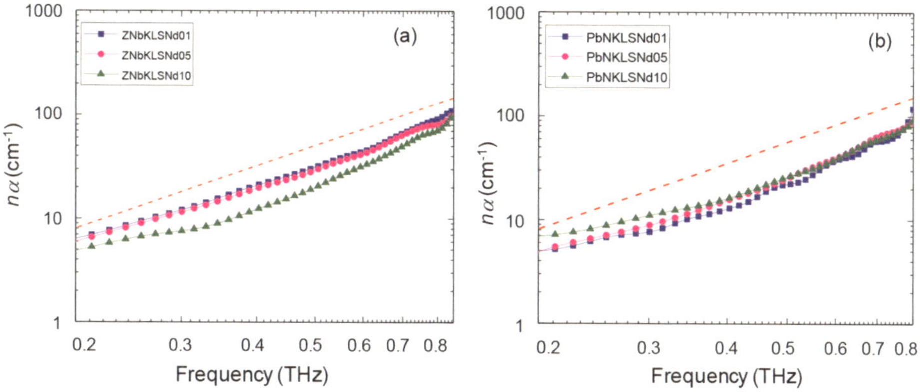

Figure 5.
Dependences of refractive index absorption coefficicent product on THz frequency for (a) ZNbKLSNd and (b) PbNKLSNd glasses with different compositions. The dotted line is a guide for eye with a slope of 2.0. (reproduced from ref. [23] with the permission of AIP Publishing).

value of β seems to be variable over a wide range ($2 \sim 3.4$ [16, 33]) depending on the alkali modifier contents in oxide glasses [33] and the coordination number in chalcogenide glasses [16]. More detailed interpretation is subject to further study.

3.2 Absorption and refractive index properties in different glasses and their physical origins

The correlation between the THz absorption parameter, K, and refractive index, n, evaluated at 0.5 THz is shown in **Figure 6** for the present OFS glasses and selected other glasses. Glass compositions corresponding to all data points in the figure can be found in [23]. Taking into consideration that the refractive indices of OFS glasses show only small changes in the measured THz frequency range, the K parameter is nearly in proportion to the absorption coefficient according to Eq. (4). The larger K implies larger absorption loss and hence suggests a glass structure involving more and/or heavier disorders. When alkali oxides such as Na_2O are introduced in the silicate oxide glass network structures, covalent Si-O-Si bonds are broken due to larger electronegativity difference between Na (0.93) and Si (1.90). This results in the introduction of local disorders in the glass structure and generates a number of non-bridging oxygens, leading to the increase of the absorption parameter as well as the refractive index [13]. Naftaly et al. [12] showed that K is proportional to the fourth power of the refractive index in a series of commercial multi-component silicate oxide glasses (except for silica), data of which are also shown in **Figure 6**. A trend of the THz absorption coefficient increasing with the THz refractive index is found in most of high-refractive index glasses. Such phenomena can be understood by considering the charge fluctuations that are generated due to the structural disorder when the glass structure is modified. Generation of such atomic charge fluctuations could be enhanced particularly in multi-component glasses such as OFS glasses. As is displayed in **Figure 6**, however, the present OFS glasses show high refractive indices yet maintaining very low absorption coefficients regardless of that their glass

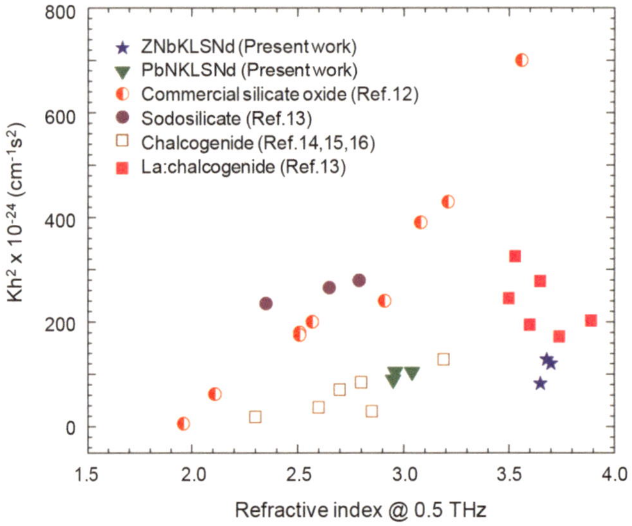

Figure 6.
Relationship between the absorption parameter and refractive index determined at 0.5 THz for a variety of multi-component glasses. (reproduced from ref. [23] with the permission of AIP Publishing).

compositions involve appreciable amounts of alkali ions (Na, K) and large atomic weight metal (Nd, Nb) oxides, which would have caused the increase of absorption losses as noticed in most of fluorine-free oxide glasses. Although some more THz-TDS based studies on silicate, germinate, tellurite oxide, and fluoride multi-component glasses have been reported very recently by Pacewicz et al. [33], the highest refractive index and lowest absorption of the present OFS glasses are still valid.

For understanding those properties of OFS glasses, it should be noticed that our OFS glasses incorporate appreciable amounts of fluorides, e.g., ZnF_2, LiF, and PbF_2. In many of oxide-based glasses, fluorine is known to restrict the formation of glass network due to its extremely high electronegativity (Pauling: 3.98) [34]. The atomic polarizability (1.04 $Å^3$) of fluorine is much lower than that of oxygen (3.88 $Å^3$) [35], and this is considered to help reduce the refractive indices of OFS glasses. On the other hand, when the glass network involves structural disorders and strains, the incorporation of fluorine can lead to their relaxation. It was found by Hosono et al. [36] that the introduction of a very small amount (>1 mol%) of fluorine into silica results in a significant enhancement of the hardness against ultraviolet radiation. According to their interpretation, the strong F-Si couplings cause the reconstruction of strained three-/four-membered Si-O-Si rings, making the network bonds more open and eventually relaxing physical disorders in the glass structure. It has also been reported that the fluorine substitutes randomly into the oxygen sites [37], and this works favorably to give rise to a uniform structural relaxation. Very low absorption characteristics as found in the present OFS glasses support that the charge fluctuation is considerably prevented in the glass structure owing to the structural relaxation effect of fluorine.

As shown in **Figure 4**, the dependence of the THz absorption coefficient of OFS glasses on the x value is weak. Dependence of the refractive index on x is even weaker. This would suggest that significantly low absorption properties of OFS glasses is predominantly determined by the basic composition of the present OFS glasses, which contain at least 10 mol% of fluorine, rather than by the incorporation of rare earth (Nd) modifier. This is consistent with prior works in which the structural relaxation is observed on the fluorine addition as low as 1 mol% in silica [36, 38]. It can be concluded that the structural relaxation effect of fluorine is primarily responsible for the significantly low absorption loss properties of the present OFS glass systems.

3.3 Dielectric properties and polarizabilities

In order to compare dielectric properties of the present OFS glasses with those of other glasses, we analyze the data by using the Clausius-Mossotti equation [12, 23, 35]. The dielectric constant in the optical frequency range, $\varepsilon_{\mathrm{opt}}$, is determined by the polarizability of electrons associated with constituent molecules in the glass. On the other hand, the dielectric constant in the sub-THz frequency range, $\varepsilon_{\mathrm{THz}}$, is determined by the total polarizability including contributions of both the fast response electrons (P_e) and the slow response ions (P_i) involved in the glass material. Both of the dielectric constants, $\varepsilon_{\mathrm{opt}}$ and $\varepsilon_{\mathrm{THz}}$, are correlated with the molar electronic polarizability, P_e, molar ionic polarizability, P_i, and molar total polarizability, $P_{tot} = P_e + P_i$, as expressed by the following formula:

$$\frac{\varepsilon_{opt} - 1}{\varepsilon_{opt} + 2} V_m = \frac{4\pi}{3} N_A P_e = R_m^e, \qquad (5)$$

$$\frac{\varepsilon_{THz}-1}{\varepsilon_{THz}+2}V_m = \frac{4\pi}{3}N_A(P_e+P_i) = R_m^{tot} = R_m^e(1+a), \tag{6}$$

where V_m is the molar volume which is defined by M/ρ using the density ρ and molecular weight M, R_m^e is the molar electronic refraction, R_m^{tot} is the molar total refraction including both electronic and ionic contributions, a is defined as the ratio P_i/P_e, and N_A is the Avogadro number (6.023×10^{23} mol^{-1}). Considering that the extinction coefficient ($k = \alpha c/(4\pi\nu)$) is negligible in comparison with the refractive index in the sub-THz region in the present glasses, the THz dielectric constant is obtained from $\varepsilon_{THz} = {n_{THz}}^2$. Similarly the optical dielectric constant is obtained from $\varepsilon_{opt} = {n_{opt}}^2$. By using Eqs. (5) and (6), the polarizabilities can be evaluated from the measured refractive indices in each frequency range. The THz refractive indices have been determined at a fixed frequency of 0.5 THz for all glasses in this study. The optical refractive indices of OFS glasses have been determined at the wavelength of 1500 nm from the measured optical reflectivity spectra using Eq. (3). For other commercial glasses, the refractive indices determined at 589 nm [12] have been used. The resultant values of polarizabilities P_e, P_i and P_{tot}, and $a = P_i/P_e$ are summarized in **Table 2**.

Glass	n_{THz}	n_{opt}	P_{tot} (Å^3)	P_e (Å^3)	P_i (Å^3)	$a = P_i/P_e$	ε_{THz}- ε_{opt}	λ_0 (μm)	g (10^{10} cm/mol)	**Ref.**
ZNbKLSNd01	3.68	1.81	10.9	5.82	5.05	0.868	10.27	74.57	6.247	[24]
ZNbKLSNd05	3.70	1.79	11.7	5.99	5.67	0.947	10.49	78.89	6.066	[24]
ZNbKLSNd10	3.65	1.80	12.6	6.83	5.80	0.850	10.08	98.72	4.186	[24]
PbNKLSNd01	2.95	1.57	8.47	3.79	4.68	1.24	6.24	126.5	1.279	[24]
PbNKLSNd05	3.04	1.63	8.77	4.19	4.58	1.10	6.58	112.6	1.510	[24]
PbNKLSNd10	2.96	1.53	9.38	3.94	5.44	1.38	6.42	115.5	1.617	[24]
Silica	1.96	1.46	5.26	2.91	2.35	0.810	1.71			[12, 24]
Pyrex	2.11	1.47	5.90	3.02	2.88	0.950	2.27			[12, 24]
BK7	2.51	1.52	6.60	3.09	3.51	1.15	3.99	80.84	1.586	[12, 24]
SK10	2.91	1.62	8.71	4.29	4.42	1.03	5.85	126.7	1.082	[12, 24]
SF10	3.21	1.73	10.7	5.66	5.07	0.900	7.31	138.7	1.305	[12, 24]
SF6	3.56	1.81	10.8	5.86	4.93	0.840	9.40	132.5	1.762	[12, 24]
S1	3.89	2.37	4.60	3.38	1.22	0.361	9.52	108.8	1.110	[16, 24]
S2	3.74	2.38	4.20	3.07	1.13	0.368	8.32	110.4	0.953	[16, 24]
S3	3.60	2.37	4.30	3.22	1.08	0.335	7.34	95.68	1.064	[16, 24]
S4	3.50	2.31	4.80	3.53	1.27	0.360	6.91	91.41	1.343	[16, 24]
S5	3.65	2.37	4.90	3.65	1.25	0.343	7.71	93.09	1.343	[16, 24]
S6	2.85	2.73	4.60	4.36	0.24	0.055	0.67	139.4	0.303	[16, 24]
S7	3.19	2.60	5.60	4.79	0.81	0.169	3.42	139.7	0.052	[16, 24]

Table 2.
Material, THz, and optical parameters determined for a variety of glasses (reproduced from ref. [23] with the permission of AIP publishing).

By using this ratio a, the electronic molar refraction R_m^e and the molar volume V_m, the optical refractive index (n_{opt}) and THz refractive index (n_{THz}) can be expressed in the following forms:

$$n_{opt} = \sqrt{\frac{1 + 2R_m^e/V_m}{1 - R_m^e/V_m}}, \tag{7}$$

$$n_{THz} = \sqrt{\frac{1 + 2R_m^e(1+a)/V_m}{1 - R_m^e(1+a)/V_m}}. \tag{8}$$

Figure 7 shows the relationships between the THz refractive index and optical refractive index for various silicate oxide glasses including OFS glasses and La:chalcogenide glasses. Two curves indicated in the figure have been obtained from Eqs. (7) and (8) using R_m^e/V_m as a variable parameter for fixed values of a; a = 0.95 and a = 0.35 have been assumed for silicate oxides and La:chalcogenides, respectively, in **Figure 7**. As shown in this figure, two different glass groups are distinguished by the polarizability ratio a. Small shifts of the measured data from the calculated curve are due to the discrepancy of the real value of a from the fixed value used for calculation, for example, a = 0.85–0.95 for ZNbKLSNd glasses and a = 1.10–1.38 for PbNKLSNd glasses, as can be confirmed in **Table 2**. A similar n_{opt} vs. n_{THz} relationship is found for La:chalcogenide glass group. On the basis of the present result, the polarizability ratio a can be a useful parameter to characterize the chemical composition/structure of glass materials at least within the same glass family such as silicate oxide glasses, in which the common oxygen ion O^{2-} is responsible to control the dielectric property.

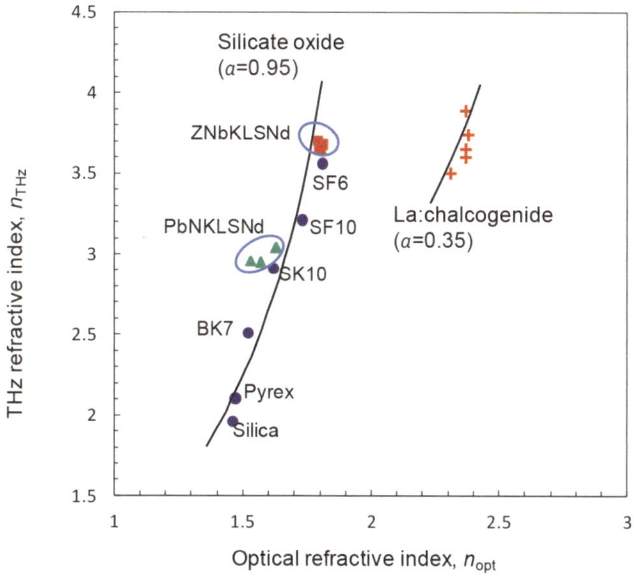

Figure 7.
Relationship of THz refractive index as a function of optical refractive index for various glasses. Curves shown have been calculated with representative values of $a = P_i/P_e$. *(ref. [24]).*

3.4 Ionic contribution to dielectric properties of silicate oxide glasses

We have so far seen that the electronic and ionic contributions to the dielectric and absorption properties of glasses are interpreted by a unified relationship for all silicate oxide glasses. In the following, we focus on the contribution of the ionic vibrations to the dielectric properties of glasses by using a simplified dielectric model [24] and discuss their relation with the physical properties of glasses. The dielectric constant (ε) at a given frequency (ν) in the THz and far-infrared (FIR) region, ε_ν, can be represented by the classical single harmonic oscillator model (Lorentz model), which describes the displacement of a pair of anion and cation, as is expressed by [39, 40]:

$$\varepsilon_\nu = \varepsilon_\infty + \frac{4\pi Ne^2}{\mu}\frac{f}{\nu_0^2 - \nu^2 - i\gamma\nu}, \tag{9}$$

where ε_∞ is the dielectric constant at the infinite frequency ($\nu = \infty$), N is the number of unit cells involving ion pairs per unit volume ($= N_A/V_m$), e is the electronic charge, μ is the reduced mass of ions given by $1/\mu = 1/m_+ + 1/m_-$, where m_+ and m_- are the masses of anion and cation, ν_0 is the characteristic resonance frequency of the oscillator, f is the oscillator strength, and γ is the damping factor for displacement. Here, single charge anion-cation pairs are assumed for simplicity; that is, a single parameter ν_0 will represent the oscillator resonance frequency and eventually control the THz dielectric characteristics of the glass. Also when the frequency range of our interest is far from the resonance frequency (as will be confirmed later), the damping factor can be neglected in Eq. (9). For the convenience of analysis, Eq. (9) is rewritten to express the dielectric constant, ε_ν, as a function of the wavelength $\lambda(= c/\nu)$, as shown by the following form (Drude-Voigt relation [41, 42]):

$$\frac{1}{\varepsilon_\nu - \varepsilon_\infty} = \frac{\pi c^2 \mu V_m}{N_A e^2 f}\left(\frac{1}{\lambda_0^2} - \frac{1}{\lambda^2}\right), \tag{10}$$

where $N = N_A/V_m$and the oscillator resonance wavelength λ_0 have been used. When we define a material dependent amplitude factor g as:

$$g = \frac{N_A e^2 f}{\pi c^2 \mu}, \tag{11}$$

Eq. (10) can be written as:

$$\frac{1}{\varepsilon_\nu - \varepsilon_\infty} = \frac{V_m}{g}\left(\frac{1}{\lambda_0^2} - \frac{1}{\lambda^2}\right). \tag{12}$$

This indicates that a plot of $1/(\varepsilon_\nu\text{-}\varepsilon_\infty)$ as a function of $1/\lambda^2$ gives a straight line, and its slope gives the value of g through $g = V_m/$ [slope]. Considering the y-intercept $1/(\varepsilon_\nu - \varepsilon_\infty)$ of this plot for $\lambda \to \infty(\nu \to 0)$, Eq. (12) leads to the following relation:

$$\varepsilon_{\nu=0} - \varepsilon_\infty = \frac{g\lambda_0^2}{V_m}. \tag{13}$$

Therefore, [slope]/[y-intercept] provides the value of λ_0^2 for each of the glass material.

In the present analysis, the values of ε_∞ have been determined from the refractive indices measured at the optical wavelength by using the relation $\varepsilon_\infty = \varepsilon_{opt} = n_{opt}{}^2$. For $\varepsilon_{\nu=0}$, we have used the value of ε (ν = 0.2 THz) = $\varepsilon_{THz} = n_{THz}{}^2$ since ε_ν virtually saturates at 0.2 THz (**Figure 4**). In **Figure 8a**, $1/(\varepsilon_n - \varepsilon_\infty)$ is plotted as a function of $1/\lambda^2$ for ZNbKLSNdx (x = 1, 5, 10) and PbNKLSNdx (x = 1, 5, 10) glasses, and all the plots show reasonable linear dependences. From these plots, the g factor and λ_0 values have been determined, as listed in **Table 2**. The same procedure has been carried out for other silicate oxide glasses by using the data reported by Naftaly et al. [12], and the resultant plots are shown in **Figure 8b** for selected multi-component silicate oxide glasses. Some of the THz-TDS data have shown only weak frequency dependences (e.g., silica and Pyrex [12]) and those data have been omitted for the accuracy in the present analysis. The values of g and λ_0 determined for all glasses including La:chalcogenide glasses are included in **Table 2**.

Results shown in **Figure 8** indicate that THz dielectric constant characteristics of all the glasses studied here can be interpreted by a simplified form of single oscillator model. Eq. (12) indicates that the dielectric constant difference $(\varepsilon_\nu - \varepsilon_\infty)$, corresponding to the ionic vibration contribution at the frequency ν, is governed by the values of V_m, g and λ_0, which are all specific to the glass material under test. In order to observe the ionic vibration contributions to the dielectric constants in different glasses, we have plotted in **Figure 9** the dielectric constant difference $\varepsilon_{THz} - \varepsilon_{opt}$ $\varepsilon_{THz} - \varepsilon_{opt}$ as a function of $g\lambda_0{}^2/V_m$. We adopt the value of $\varepsilon(v = 0.2\text{ THz})$ as ε_{THz} for estimating $\varepsilon_{THz} - \varepsilon_{opt}$, and all other parameters are taken from the experimental data already described. **Figure 9** clearly shows a linear relationship regardless of the difference of glass materials (including La:chalcogenide glasses: plots not shown). This result confirms the wide applicability of the single oscillator model for assessing THz dielectric properties in a wide variety of glass materials.

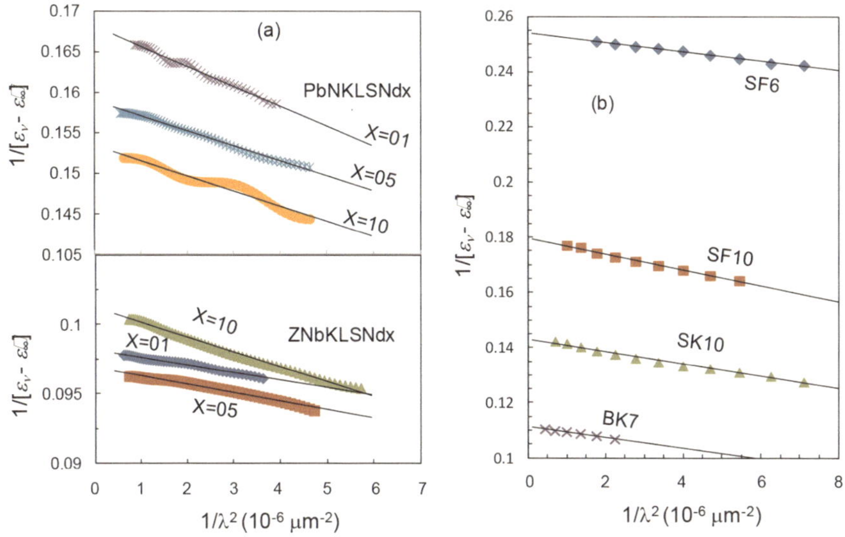

Figure 8.
$1/(\varepsilon_\nu - \varepsilon_\infty)$ versus $1/\lambda^2$ plots for (a) ZNbKLSNd and PbNKLSNd glasses with different compositions and (b) selected multi-component silicate oxide glasses. (ref. [24]).

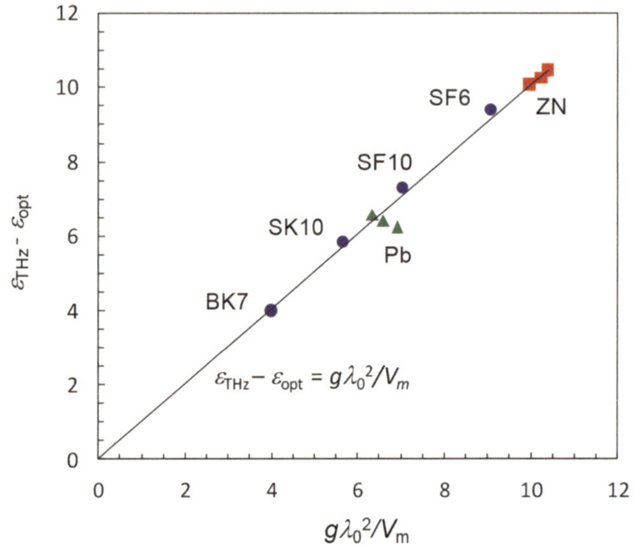

Figure 9.
Relationship between $\varepsilon_{THz} - \varepsilon_{opt}$ versus $g\lambda_o^2/V_m$ for selected silicate oxide glasses. Plots show experimental data determined from the single oscillator analysis and the straight line indicates a slope of unity. (ref. [24]).

As we have seen already, OFS glasses possess THz dielectric constants much larger than for most of other glasses, and it is very interesting and important to know the physical origin of such dielectric constant enhancement. For finding out what physical parameter controls the ionic dielectric constant, $\varepsilon_{THz} - \varepsilon_{opt}$ has been plotted as a function of each of the parameters: $1/V_m$, g and λ_0^2, as is shown in **Figure 10**. Here, $1/V_m$ is regarded as a parameter in proportion to the number of ion pairs in a unit volume. Two other parameters, g and λ_0^2, are adopted to represent the nature of vibrational oscillator. As **Figure 10a** indicates, the molar volume seems not to influence seriously the dielectric constant difference $\varepsilon_{THz} - \varepsilon_{opt}$. **Figure 10b** and **c** show

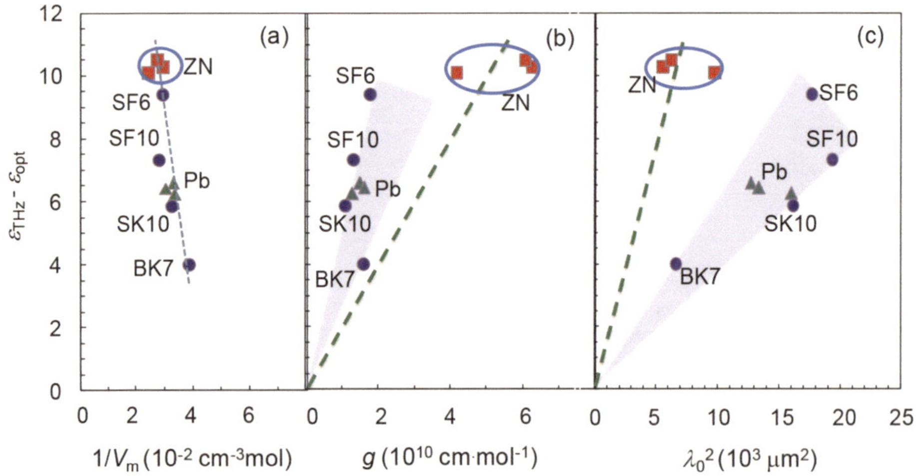

Figure 10.
The dielectric constant difference $\varepsilon_{THz} - \varepsilon_{opt}$ as functions of $1/V_m$, g and λ_0^2 for silicate oxide glasses including PbNKLSNd (Pb) and ZNbKLSNd (ZN) glasses. A significant behavior difference is found between ZNbKLSNd glasses and other silicate oxide glasses. (ref. [24]).

general trends, in which the dielectric constant difference increases in proportion to g as well as λ_0^2, supporting the trend of Eq. (13). Looking at the g and λ_0^2 dependences more closely, it is noticed that there are two distinct groups: ZNbKLSNd glasses, and all other silicate oxide glasses. In ZNbKLSNd glasses, an extremely large dielectric constant enhancement $\left(\varepsilon_{\mathrm{THz}}-\varepsilon_{\mathrm{opt}}>10\right)$ is achieved due to the significantly large value of g (rather than by λ_0^2). In the group of other silicate oxide glasses including PbNKLSNd glasses, the enhancement of the dielectric constant is caused contrastively by the large values of λ_0^2 (rather than by g). Thus, the present analysis based on the single oscillator model has elucidated that the different physical parameters are responsible for the ionic dielectric constant enhancement in different glass groups; the large g plays a principal role in ZNbKLSNd glasses and the large λ_0^2 plays leading roles in other silicate oxide glasses.

3.5 Dielectric property analysis using unified single oscillator model

3.5.1 Single oscillator-based formalism for dielectric property analysis

As we have seen in the previous section, a simplified form of single oscillator model has been found to be useful as a unified model to characterize the THz dielectric constant properties in all the glass groups studied. However, the analysis has assumed the diatomic chain vibration without considering the local field effect. To properly discuss the behavior of electromagnetic waves induced by ionic vibration, the application of phonon polariton formalism with the inclusion of the local field effect is useful [43–47]. In the following, we summarize the basic dielectric function and some useful relations for characterizing dielectric properties by referring to Gresse's approach [47]. In this section, SI unit is used for convenience, but the conversion to CGS unit can be done by replacing the vacuum dielectric constant ε_0 by $1/4\pi$.

We assume a single oscillator model based on simple diatomic chain for describing the ionic vibration to determine the dielectric functions of glass materials. An equation of motion which describes the relative displacement of the cation-anion pair $\mathbf{u}$ is written in the following form:

$$\mu\ddot{\mathbf{u}}+\mu\gamma\dot{\mathbf{u}}+D\mathbf{u}=q^{*}\mathbf{E}_{loc}, \tag{14}$$

where μ is the reduced mass of ions defined by $1/\mu=1/m_{+}+1/m_{-}$, where m_{+} and m_{-} are the anion and cation masses, γ is the damping factor for displacement, D is the force constant of the ionic pair, and q^{*} is the effective charge of ion. $\mathbf{E}_{loc}$ stands for the local electric field involving the microscopic Lorenz field applied to an atom. This local field is correlated with the macroscopic applied field $\mathbf{E}_a$ and polarization $\mathbf{P}$ by the following equation:

$$\mathbf{E}_{loc}=\mathbf{E}_a+\frac{1}{3}\cdot\frac{\mathbf{P}}{\varepsilon_0}=\left(\frac{1}{\chi}+\frac{1}{3}\right)\frac{\mathbf{P}}{\varepsilon_0}, \tag{15}$$

where a macroscopic relation of the applied field $\mathbf{E}_a$ with the polarization, $\mathbf{P}=\varepsilon_0\chi\mathbf{E}_a$, has been used. Here, ε_0 and χ are the vacuum dielectric constant and macroscopic electric susceptibility of the material. The total polarization $\mathbf{P}$ consists of the lattice strain $(\mathbf{u})$ induced dipole moment $(\mathbf{P}_{PH})$, which is originated from the relative ion core displacement, and the electronic polarization $(\mathbf{P}_{VE})$, which is caused by the valence electron displacement with respect to ion core in the local field. This correlation is expressed by:

$$\mathbf{P} = \mathbf{P}_{PH} + \mathbf{P}_{VE} = Nq^*\mathbf{u} + N\varepsilon_0 P_e \mathbf{E}_{loc}, \tag{16}$$

where N is the number of atoms in unit molecule ($N = N_\mathrm{A}/V_\mathrm{m}$) and $P_\mathrm{e}$is the electronic polarizability of valence electrons surrounding the steady ion core (without lattice strain). The influence of valence electron displacement which is caused by the lattice strain is taken into account through the effective charge q^* of ion pair. Applying the form of harmonic electric field ($\mathbf{E}_\mathrm{a} \propto \exp(-i\omega t)$), Eq. (14) can be written as:

$$\left(\omega^2 + i\omega\gamma - \omega_0{}^2\right)\mathbf{u} + \frac{q^*}{\mu}\mathbf{E}_{loc} = 0, \tag{17}$$

where ω_0 is defined as $\omega_0{}^2 = D/\mu$ by using the force constant D, and ω_0 indicates the intrinsic resonance frequency for the undamped vibration of diatomic chain under zero electric field: $\ddot{\mathbf{u}} + \omega_0{}^2\mathbf{u} = 0$.

Eqs. (15)–(17) form a set of homogeneous equations for $\mathbf{u}$, $\mathbf{P}$ and $\mathbf{E}_{loc}$ (Huang-Szigetti equation [47]). By solving the secular equation for the macroscopic susceptibility χ, the following relation is obtained:

$$\chi = \frac{3(NP_e + \chi_0)}{3 - (NP_e + \chi_0)}. \tag{18}$$

Here, NP_e and χ_0 indicate the microscopic electronic and ionic susceptibilities. The microscopic ionic susceptibility, χ_0, is induced by the ionic vibration, in which effects of valence electron polarization ($N\varepsilon_0 P_\mathrm{e}\mathbf{E}_{loc}$) and the induced local field ($\mathbf{P}/3\varepsilon_0$) are excluded, and is expressed by:

$$\chi_0 = \frac{\omega_p{}^2}{\omega_0{}^2} \cdot \frac{\omega_0{}^2}{\omega_0{}^2 - \omega^2 - i\omega\gamma}. \tag{19}$$

where ω_p is the ionic vibration resonance frequency which is defined by $\omega_p{}^2 = \frac{Nq^{*2}}{\varepsilon_0\mu}$. With a slight manipulation of Eq. (18) using the relation of $\varepsilon = 1 + \chi$, it is readily shown that the dielectric constant at the frequency ω, $\varepsilon(\omega)$, can be expressed by the following simple form equivalent to Eq. (9):

$$\varepsilon(\omega) = \varepsilon_\infty + (\varepsilon_s - \varepsilon_\infty)\frac{\omega_T{}^2}{\omega_T{}^2 - \omega^2 - i\omega\gamma}, \tag{20}$$

where ε_s and ε_∞ are the values of ε at the frequency $\omega \to 0$ and $\omega \to \infty$, respectively, and ω_T is the renormalized resonant frequency which is lowered from ω_0 due to the local field effect as is expressed by:

$$\omega_T{}^2 = \omega_0{}^2 - \frac{1}{3 - NP_e}\omega_p{}^2 = \omega_0{}^2 - \frac{1}{3}\omega_p{}^2\frac{\varepsilon_\infty + 2}{3}. \tag{21}$$

To distinguish the effects of the ionic vibration ($\mathbf{P}_{PH}$) and valence electron displacement ($\mathbf{P}_{VE}$) contributions in the determination of the overall polarization, it is necessary to evaluate the microscopic electronic and ionic susceptibility components,

NP_e and $(\omega_p/\omega_0)^2$, respectively (refer to Eqs. (18) and (19)). Since those quantities cannot be directly measured, it would be convenient if they could be monitored by using the measurable physical parameters such as the dielectric constants, ε_s and ε_∞, and the renormalized characteristic frequency, ω_T, etc. Interestingly, this is made possible by using the following equations, which are deduced from Eqs. (18)–(21):

$$\varepsilon_\infty = \frac{3 + 2NP_e}{3 - NP_e}, \tag{22}$$

$$\varepsilon_s = \frac{3 + 2\left((NP_e + (\omega_p/\omega_0)^2\right)}{3 - \left((NP_e + (\omega_p/\omega_0)^2\right)}, \tag{23}$$

$$\frac{\omega_T}{\omega_0} = \left(1 - \frac{1}{3 - NP_e}\left(\frac{\omega_p}{\omega_0}\right)^2\right)^{1/2}. \tag{24}$$

It would be noteworthy that we can also deduce from Eqs. (18)–(21) very simple and convenient relations to determine NP_e, ω_p/ω_0 and ω_T/ω_0 and a parameter by using only the steady state values of dielectric constants, ε_s and ε_∞, as shown in the following:

$$NP_e = \frac{3(\varepsilon_\infty - 1)}{\varepsilon_\infty + 2}, \tag{25}$$

$$\frac{\omega_p}{\omega_0} = \left(\frac{9(\varepsilon_s - \varepsilon_\infty)}{(\varepsilon_s + 2)(\varepsilon_\infty + 2)}\right)^{1/2}, \tag{26}$$

$$\frac{\omega_T}{\omega_0} = \left(\frac{\varepsilon_\infty + 2}{\varepsilon_s + 2}\right)^{1/2}, \tag{27}$$

$$a = \frac{P_i}{P_e} = \frac{3(\varepsilon_s - \varepsilon_\infty)}{(\varepsilon_s + 2)(\varepsilon_\infty - 1)}. \tag{28}$$

Considering $\varepsilon_\infty = \varepsilon_{opt}$ and $\varepsilon_s = \varepsilon_{THz}$, all those characteristic parameters are readily determined by only using the dielectric constants obtained in the optical and THz spectroscopic measurements. This feature is particularly advantageous as a unified model for the assessment of physical properties of different glass materials.

3.5.2 Comparison between theory and experiment

We analyze the measured dielectric properties of different glasses by using the relations described above. The renormalized characteristic frequency ω_T corresponds to the characteristic resonance wavelength λ_0 as used in Section **3.4** through $\omega_T = 2\pi c/\lambda_0$. All the characteristic parameters, NP_e, ω_p/ω_0 and ω_T/ω_0 and a, are obtained by using Eqs. (25)–(28) and the result is summarized in **Table 3**.

In **Figure 11**, a set of three-dimensional maps are shown to visualize the behaviors of (a) ε_s, (c) ω_T/ω_0, (e) ε_∞, and (f) a parameter on the planes of NP_e and ω_p/ω_0. **Figure 11** (**b**) and (**d**) show the values of (**b**) ε_s and (**d**) ω_T/ω_0 which have been obtained from the experimental data of ε_s and ε_∞. As is realized in **Figure 11e** and also in Eq. (22), ε_∞ is independent from the ionic contribution and is determined solely by

Glass	ε_{THz}	ε_{opt}	a	ω_T^2 (x10^2 THz^2)	ε_s - ε_∞	NP_e	ω_p/ω_0	ω_T/ω_0	I_P	Ref
ZNbKLSN01	13.54	3.27	0.868	6.39	10.26	1.294	1.061	0.583	0.465	[25]
ZNbKLSN05	13.69	3.20	0.947	5.71	10.49	1.271	1.075	0.576	0.476	[25]
ZNbKLSN10	13.32	3.24	0.850	3.65	10.08	1.282	1.063	0.585	0.468	[25]
PbNKLSN01	8.70	2.46	1.23	2.22	6.238	0.984	1.084	0.646	0.544	[25]
PbNKLSN05	9.24	2.66	1.09	2.80	6.585	1.067	1.064	0.644	0.515	[25]
PbNKLSN10	8.76	2.34	1.38	2.66	6.421	0.927	1.112	0.635	0.572	[25]
Silica	3.84	2.13	0.808		1.710	0.822	0.799	0.841	0.437	[12, 25]
Pyrex	4.43	2.16	0.954		2.270	0.837	0.874	0.804	0.477	[12, 25]
BK7	6.30	2.31	1.14	5.44	3.990	0.912	1.002	0.721	0.524	[12, 25]
SK10	8.47	2.62	1.03	2.21	5.844	1.053	1.042	0.665	0.508	[12, 25]
SF10	10.30	2.99	0.896	1.85	7.311	1.197	1.035	0.637	0.472	[12, 25]
SF6	12.67	3.28	0.841	2.02	9.398	1.294	1.046	0.600	0.458	[12, 25]

Table 3.
THz and optical dielectric parameters determined for a variety of glasses (ref. [25]).

the polarization due to the valence electron contribution. In contrast, ε_s is controlled by both the electron and ion contributions, and it increases rapidly as those contributions become larger, as is displayed in **Figure 11a**. The contours delineated for $NP_e < 3$ and $\omega_p/\omega_0 < 3^{1/2}$ on the base planes in **Figure 11b** and **d** indicate the limits of existence of ε_s where the catastrophic collapse of glass phase occurs and the dielectric constant diverges to infinity. In **Figure 11b**, measured data of ε_s are displayed for various silicate oxide glasses and also for two La: chalcogenide glasses (S1 and S5 in **Table 2**) [16, 24, 25]. As for OFS glasses, only data for x = 5 are shown for avoiding complexity of the figure. The comparison between figures (a) and (b) indicates a good correspondence between the experimental and calculated results for all glasses. More precise comparison has been reported in [25]. Since chalcogenide glasses are known to be highly covalent in comparison with silicate oxide glasses, they appear at locations with higher NP_e and lower ω_p/ω_0 than for silicate oxide glasses.

In **Figure 11c** and **d**, calculated and measured values of ω_T/ω_0 are shown (for OFS glasses, x = 5 only) with a different viewing angle for the ease of observing the overall trend. It is clearly seen that as the electronic and ionic contributions increase, the characteristic frequency exhibits a severe softening (lowering) toward the catastrophic boundary where ω_T becomes zero and ε_s diverges to infinity. As shown in those figures, the experimental data support the calculated trend. This softening behavior of ω_T/ω_0 corresponds to the diverging increase of the THz dielectric constant, and this feature can be used for detecting the characteristic change of dielectric property. **Figure 11f** shows the behavior of a parameter as evaluated by Eq. (28), which exhibits a non-monotonous variation in contrast to the behaviors of ε_s andε_∞.

Those results suggest that the dielectric parameters in the glasses examined here are influenced primarily by the ionic vibration originated polarization $\left((\omega_p/\omega_0)^2\right)$ but precise parameter values are modified by the valence electron polarization effect

DOI: http://dx.doi.org/10.5772/intechopen.111635

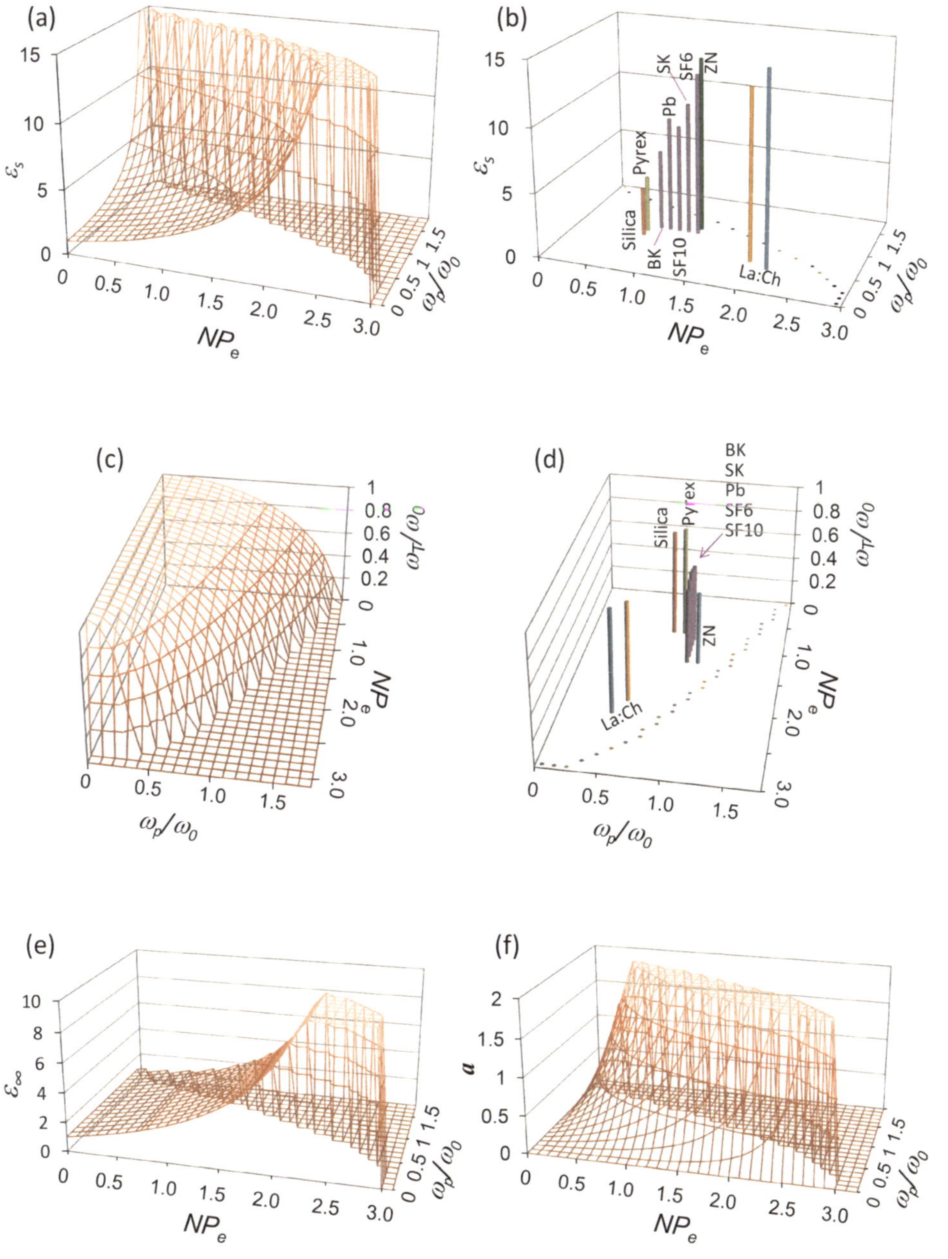

Figure 11.
A variety of the dielectric parameters (a), (b): ε_s, (c), (d): ω_T/ω_0, (e): ε_∞, and (f): $a = P_i/P_e$ plotted on the base planes of NP_e and ω_p/ω_0. Plots in (a), (c), (e), and (f) show the results calculated by using Eqs. (22)–(24), and plots in (b) and (d) show measured ε_s and ω_T/ω_0 evaluated from measured ε_s and ε_∞ values using Eqs. (25)–(28). (ref. [25]).

(NP_e). This explains the reason of data scattering in the ε_s versus ε_∞ relationship as noticed in **Figure 7**, and accurate values of NP_e and $(\omega_p/\omega_0)^2$ should be taken into account for exact fitting. Precise comparison has confirmed the consistency between the model and experiment, as has been described in Ref. [25]. This indicates that values of all key parameters necessary for the analysis can be determined simply from Eqs. (25)–(28).

Here, we look at the behavior of a parameter in more detail. As is noticed in **Figure 11f**, the envelope of a parameter variation drawn on the $NP_e - \omega_p/\omega_0$ plane is warped differently when compared with the envelopes of ε_∞ (**Figure 11e**) and ε_s (**Figure 11a**). Actually in **Figure 7**, the a parameter value, which is required for fitting the Clausius-Mossotti curves and measured ε_{opt} and ε_{THz} data, exhibits a fairly complicated behavior (e.g., double-valued ε_{opt} for the same a; refer to **Table 2**) as already mentioned above. Such a non-monotonous behavior of a is readily understood by taking account of the difference in the envelope warp for a and ε_∞: when NP_e increases, ε_∞ also increases, whereas a parameter can either increase or decrease depending on the value of ω_p/ω_0.

We define here the polarization ionicity I_P which is expressed by:

$$I_P = \frac{P_i}{P_e + P_i} = \frac{(\omega_p/\omega_0)^2}{NP_e + (\omega_p/\omega_0)^2} = \frac{3(\varepsilon_s - \varepsilon_\infty)}{(\varepsilon_s - 1)(\varepsilon_\infty + 2)} = \frac{a}{1+a}, \tag{29}$$

where Eqs. (22)–(28) have been used. The values of I_P evaluated for different glasses are included in **Table 3**. **Figure 12** shows the overview of I_P behavior in the $NP_e - \omega_p/\omega_0 - I_P$ space, in which I_P exhibits monotonous variation and a diverging behavior as seen for the a parameter (**Figure 11f**) is eliminated. Because of this nature, although both parameters, a and I_P, represent essentially the degree of ionicity, I_P is considered to be more useful for characterizing the ionic contribution in the low-frequency dielectric constant of materials. For example, the discussion in Section **3.3** can be rephrased by using I_P instead of $a(= I_P/(1 - I_P))$ so that the ionicity effect is expressed more explicitly.

3.5.3 Assessment of dielectric constant enhancement in OFS glasses

We have shown earlier in **Figure 10** that ZNbKLSNd glass shows appreciably higher g factor and significantly smaller λ_0^2 in comparison with those of most all other silicate oxide glasses. However, g factor in Eq. (9) corresponds to $(\varepsilon_s - \varepsilon_\infty)\omega_T{}^2$ in Eq. (20) and

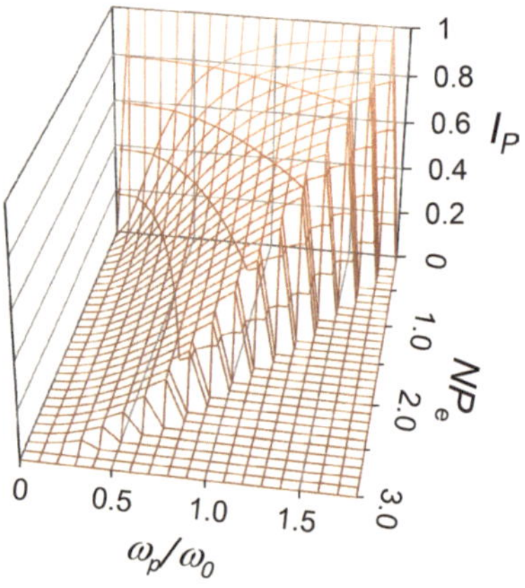

Figure 12.
Three dimensional view of the behavior of the ionicity contribution (I_P) to low-frequency dielectric constant in the plane of NP_e and ω_p/ω_0. (ref. [25]).

this involves the contributions of both electronic (NP_e) and ionic (ω_p/ω_0) contributions, as is shown in **Figure 11a** and **Figure 11b**. Similarly, λ_0^2 corresponds to $1/\omega_T{}^2$, and this also involves both contributions. In contrast, it is indicated in **Figure 11** that as the electronic and ionic susceptibility amplitudes increase, the THz dielectric constant is enhanced and the softening of the characteristic resonant frequency becomes severer. These effects are strengthened significantly as the total susceptibility $\left(NP_e + (\omega_p/\omega_0)^2\right)$ comes closer to the catastrophic limit equal to 3, at which ε_∞ diverges to infinity and ω_T/ω_0 becomes zero. This unique feature is very useful for distinguishing the mechanism of dielectric constant enhancement in different glasses.

Large ε_s enhancement in ZNbKLSNd glass is primarily due to the significantly large electronic polarizability together with a reasonably high ionicity, which makes its location in NP_e-ω_{T}/ω_0 plane considerably close to the catastrophic boundary. On the other hand, in the case of PbNKLSNd glasses, although the ionicity ($I_P = 0.515 - 0.572$) is much higher than for ZNbKLSNd glass ($I_P = 0.465 - 0.476$), the electronic polarizability is smaller, and the overall susceptibility is not very close to the catastrophic boundary, resulting in a limited enhancement of ε_s. Fused silica has small polarizability and small ionicity ($I_P = 0.437$), and exhibits correspondingly small ε_s. La:chalcogenide glasses (S1 and S5) included in **Figure 11** for reference are considered to be highly covalent, and their enhancement in ε_s is primarily due to their electronic polarizability effect rather than the ionicity, which is consistent with their optical dielectric constant ε_∞ much higher than other glasses (See **Table 2**). Dielectric properties of other glasses are also interpreted by referring to their locations in the NP_e-ω_T/ω_0 plane.

Thus, the interplay between the electronic polarizability and ionicity determines the dielectric constant enhancement, and this feature is clearly visualized in the present analysis based on a precise single oscillator model. Physical origin of the differences found in g and $\lambda_o{}^2$ behaviors in **Figure 11** has been clarified by this analysis. By combining these results with other microscopic/macroscopic analyses, more detailed physical/chemical properties can be explored. Characterization on some more properties such as the electronegativity [25], bandgap energy [48] and infrared absorption spectra [30] is discussed in our recent publications.

4. Conclusion

Newly developed OFS glasses have been characterized by using THz-TDS technique, and the highest THz dielectric constant (max. 13.7) and lowest THz absorption loss (min. 5.6 cm^{-1}) in a series of multi-component silicate oxide glasses have been demonstrated. It has been shown from the frequency dependence of the absorption that the structural disorder-originated charge fluctuation are suppressed in the present OFS glasses presumably due to fluorine's structural relaxing effect, resulting in the lowest absorption even for the highest dielectric constant in ZNbKLSNd glass. Analysis based on a simplified single oscillator model has shown that ZNbKLSNd glass exhibits oscillator parameters significantly different from those of PbNKLSNd glasses as well as of other silicate oxide glasses. A unified analysis method based on a more precise single oscillator dielectric model, which can distinguish the electronic and ionic contributions, has been developed and applied to further characterization. Along with this unified model, the polarization ionicity parameter has been introduced and confirmed to be a useful parameter to indicate the material's ionicity. The dielectric constant enhancement found in the silicate oxide glasses including OFS glasses has been systematically interpreted by the interplay between the electronic polarizability and material's ionicity. Unique properties of ZNbKLSNd glasses have been attributed primarily to their highest electronic polarizability. In contrast, the ionicity has been found to be more influential to enhance the dielectric constant in PbNKLSNd glasses. The present dielectric property analysis method based on THz-TDS technique provides a powerful tool for characterizing detailed physical nature of a variety of materials.

Acknowledgements

The authors are thankful to Ministry of Science and Technology (National Council of Science and Technology), Taiwan, for funding support to the present research (# MOST 106–2112- M-007–022-MY2, # MOST 110–2923-E-007–006). Doddoji Ramachari is thankful to Ministry of Science and Technology, Taiwan for the Post-doctoral Fellowship award during this research work. The authors thank Chao-Kai Wang and Chun-Ling Yen for their support in the THz-TDS measurements. Osamu Wada thanks to Japan Society for the Promotion of Science (JSPS KAKENHI Grant Number JP19K04532) for supporting this publication.

Conflict of interest

The authors declare no conflict of interest.

Author details

Osamu Wada[1,2]*, Doddoji Ramachari[3,4], Chan-Shan Yang[5,6], Takashi Uchino[7] and Ci-Ling Pan[1]

1 Department of Physics, National Tsing Hua University, Hsinchu, Taiwan

2 Office for Academic and Industrial Innovation, Kobe University, Kobe, Japan

3 Institute of Research and Development, Duy Tan University, Da Nang, Vietnam

4 Faculty of Natural Sciences, Duy Tan University, Da Nang, Vietnam

5 Institute and Undergraduate Program of Electro-Optical Engineering, National Taiwan Normal University, Taipei, Taiwan

6 Micro/Nano Device Inspection and Research Center, National Taiwan Normal University, Taipei, Taiwan

7 Graduate School of Science, Kobe University, Kobe, Japan

*Address all correspondence to: fwga3962@nifty.com

IntechOpen

© 2023 The Author(s). Licensee IntechOpen. This chapter is distributed under the terms of the Creative Commons Attribution License (http://creativecommons.org/licenses/by/3.0), which permits unrestricted use, distribution, and reproduction in any medium, provided the original work is properly cited.

References

[1] Grischkowsky D, Keiding S, van Exter M, Fattinger C. Far-infrared time-domain spectroscopy with terahertz beams of dielectrics and semiconductors. Journal of Optical Society of America B. 1990;**7**:2006-2015. DOI: 10.1364/JOSAB.7.002006

[2] Tonouchi M. Cutting-edge terahertz technology. Nature Photonics. 2007;**1**:97-105. DOI: 10.1038/nphoton.2007.3

[3] Yun-Shik Lee, Principles of Terahertz Science and Technology. New York: Springer; 2009. DOI: 10.1007/978-0-387-09540-0

[4] Zhang X-C, Xu J-Z. Introduction to THz wave photonics. New York: Springer; 2010. DOI: 10.1007/978-1-4419-0978-7

[5] Kai-Erik Peiponen J, Zeitler A, Kuwata-Gonokami M, editors. Terahertz Spectroscopy and Imaging. Heidelberg: Springer; 2013. DOI: 10.1007/978-3-642-29564-5

[6] Bolivar PH, Brucherseifer M, Rivas JG, Gonzalo R, Ederra I, Reynolds AL, et al. Measurement of the dielectric constant and loss tangent of high dielectric-constant materials at terahertz frequencies. IEEE Transactions on Microwave Theory and Techniques. 2003;**51**:1062. DOI: 10.1109/TMTT.2003.809693

[7] Yang C-S, Chang C-M, Chen P-H, Yu P, Pan C-L. Broadband terahertz conductivity and optical transmission of indium-tin-oxide (ITO) nanomaterials. Optics Express. 2013;**21**:16670. DOI: 10.1364/OE.21.016670

[8] Beard MC, Turner GM, Schmuttenmaer CA. Subpicosecond carrier dynamics in low-temperature grown GaAs as measured by time-resolved terahertz spectroscopy. Journal of Applied Physics. 2001;**90**:5915. DOI: 10.1063/1.1416140

[9] Naftaly M, Miles RE. Terahertz time-domain spectroscopy for material characterization. Proceedings of IEEE. 2007;**95**:1658. DOI: 10.1109/JPROC.2007.898835

[10] Yang C-S, Lin C-J, Pan R-P, Que CT, Yamamoto K, Tani M, et al. The complex refractive indices of the liquid crystal mixture E7 in the terahertz frequency range. Journal of Optical Society of America B. 2010;**27**:1866. DOI: 10.1364/JOSAB.27.001866

[11] Jepsen PU, Cooke DG, Koch M. Terahertz spectroscopy and imaging – Modern techniques and applications. Laser Photonics Reviews. 2011;**5**:124. DOI: 10.1002/lpor.201000011

[12] Naftaly M, Miles RE. Terahertz time-domain spectroscopy of silicate glasses and the relationship to material properties. Journal of Applied Physics. 2007;**102**(4):043517/1-6. DOI: 10.1063/1.2771049,043517/1-6

[13] Parrott EPJ, Axel J, Zeitler G, Simon B, Hehlen LF, Gladden SN, et al. Atomic charge distribution in sodosilicate glasses from terahertz time-domain spectroscopy. Physical Review B. 2010;**82**(14):140203/1-4. DOI: 10.1103/PhysRevB.82.140203

[14] Kang SB, Kwak MH, Park BJ, Kim S, Ryu H-C, Chung DC, et al. Optical and dielectric properties of chalcogenide glasses at terahertz frequencies. ETRI Journal. 2009;**31**(6):667-674. DOI: 10.4218/etrij.09.1209.0028

[15] Zalkovskij M, Bisgaard CZ, Novitsky A, Malureanu R, Savastru D. Ultrabroadband terahertz spectroscopy of chalcogenide glasses. Applies Physics Letters. 2012;**100**(3):031901/1-4. DOI: 10.1063/1.3676443

[16] Ravagli A, Naftaly M, Craig C, Weatherby E, Hewak DW. Dielectric and structural characterisation of chalcogenide glasses via terahertz time-domain spectroscopy. Optical Materials. 2017;**69**:339-343. DOI: 10.1016/j.optmat.2017.04.057

[17] Jha A. Inorganic Glasses for Photonics: Fundamentals, Engineering and Applications. West Sussex: John Wiley & Sons; 2016. DOI: 10.1002/9781118696088

[18] Tanaka K, Shimakawa K. Amorphous Chalcogenide Semiconductors and Related Materials. New York: Springer; 2011. DOI: 10.1007/978-1-4419-9510-0

[19] Ramachari D, Moorthy LR, Jayasankar CK. Optical absorption and emission properties of Nd^{3+}-doped oxyfluorosilicate glasses for solid state lasers. Infrared Physics and Technology. 2014;**67**:555-559. DOI: 10.1016/j.infrared.2014.09.020

[20] Viswanath CSD, Krishnaiah KV, Jayasankar CK. Luminescence properties of europium doped oxyfluorosilicate glasses for visible light devices. Optical Materials. 2018;**83**:348. DOI: 10.1016/j.optmat.2018.05.057

[21] Manasa P, Ramachari D, Kaewkhao J, Meejitpaisan P, Kaewnuam E, Joshi AS, et al. Studies of radiative and mechanical properties of Nd^{3+}-doped lead fluorosilicate glasses for broadband amplification in a chirped pulse amplification based high power laser system. Journal of Luminescence. 2017;**188**:558-566. DOI: 10.1016/j.jlumin.2017.04.065

[22] Zalkovskij M, Strikwerda AC, Iwaszczuk K, Popescu A, Savastru D, Malureanu R, et al. Terahertz-induced Kerr effect in amorphous chalcogenide glasses. Applied Physics Letters. 2013;**103**:221102. DOI: 10.1063/1.4832825

[23] Ramachari D, Yang C-S, Wada O, Uchino T, Pan C-L. High-refractive index, low-loss oxyfluorosilicate glasses for sub-THz and millimeter wave applications. Journal of Applied Physics. 2019;**125**(15):151609/1-8. DOI: 10.1063/1.5083091

[24] Wada O, Ramachari D, Yang C-S, Uchino T, Pan C-L. High refractive index properties of oxyfluorosilicate glasses and a unified dielectric model of silicate oxide glasses in sub-terahertz frequency region. Optical Materials Express. 2020;**10**(2):607-621. DOI: 10.1364/OME.382686

[25] Wada O, Ramachari D, Yang C-S, Uchino T, Pan C-L. Systematic characterization of THz dielectric properties of multi-component glasses using unified oscillator model. Optical Materials Express. 2021;**11**(3):858-874. DOI: 10.1364/OME.417771

[26] Yang C-S, Lin M-H, Chang C-H, Yu P, Shieh J-M, Shen C-H, et al. Non-drude behavior in indium-tin-oxide nanowhiskers and thin films investigated by transmission and reflection THz time-domain spectroscopy. IEEE Journal of Quantum Electronics. 2013;**49**(8):677-690. DOI: 10.1109/JQE. 2013.2270552

[27] Parrott EPJ, Fischer BM, Gladden LF, Zeitler JA, Jepsen PU. Terahertz spectroscopy of crystalline and non-crystalline solids, chap. 8. In: Kai-Erik Peiponen J, Zeitler A, Kuwata-

Gonokami M, editors. Terahertz Spectroscopy and Imaging. Heidelberg: Springer; 2013. DOI: 10.1007/978-3-642-29564-5

[28] Roux J-F, Garet F, Coutaz J-L. Principles and applications of THz time domain spectroscopy, chap. 8. In: Perenzoni M, Paul DJ, editors. Physics and Applications of Terahertz Radiation. Dordrecht: Springer; 2014. DOI: 10.1007/978-94-007-3837-9

[29] Kuzmany H. Solid-State Spectroscopy - an Introduction. Berlin: Springer; 1998. DOI: 10.1007/978-3-662-03594-8

[30] Wada O, Ramachari D, Yang C-S, Harada Y, Uchino T, Pan C-L. Mechanism of THz dielectric constant enhancement in multi-component oxide glasses investigated by infrared and THz spectroscopies. Journal of Physics and Chemistry of Solids. 2023; **176**:111237. DOI: 10.1016/j.jpcs.2023.111237

[31] Strom U, Hendrickson JR, Wagner RI, Taylor PC. Disorder-induced far infrared absorption in amorphous materials. Solid State Communications. 1974;**15**:1871. DOI: 10.1016/0038-1098(74)90106-9

[32] Strom U, Taylor PC. Temperature and frequency dependences of the far-infrared and microwave optical absorption in amorphous materials. Physical Review B. 1977;**16**:5512. DOI: 10.1103/PhysRevB.16.5512

[33] Pacewicz A, Cimek J, Salki B, Walczakowski M, Buczynski R. Reconstruction and modeling of the complex refractive index of nonlinear glasses from terahertz to optical frequencies. Optics Express. 2021;**29**(16): 26191. DOI: 10.1364/OE.431430

[34] WebElements [Internet] Available from: https://www.webelements.com [Accessed: February 20, 2023]

[35] Kittel C. Introduction to Solid State Physics. 8th ed. New York: John Wiley and Sons; 2005. Chap. 16

[36] Hosono H, Mizuguchi M, Skuja L. Fluorine-doped SiO_2 glasses for F_2 excimer laser optics: Fluorine content and color-center formation. Optics Letters. 1999;**24**:1549. DOI: 10.1364/OL.24.001549

[37] Duncan TM, Douglass DC, Csencsits R, Walker KL. Study of fluorine in silicate glass with ^{19}F nuclear magnetic resonance spectroscopy. Journal of Applied Physics. 1986;**60**:130. DOI: 10.1063/1.337675

[38] Funabiki F, Kamiya T, Hosono H. Doping effects in amorphous oxides. Journal of Ceramic Society of Japan. 2012;**120**:447. DOI: 10.2109/jcersj2.120.447

[39] Wooten F. Optical Properties of Solids. New York: Academic Press; 1972. DOI: 10.11316/butsuri1946.28.9.803

[40] Fox M. Optical Properties of Solids. New York: Oxford University Press; 2001

[41] Wood RW. Physical Optics. 3rd ed. New York: Macmillan; 1934. p. 486

[42] Takebe H, Fujino S, Morinaga K. Refractive-index dispersion of tellurite glasses in the region from 0.40 to 1.71 μm. Journal of the American Ceramic Society. 1994;**77**(9):2455-2457. DOI: 10.1111/j.1151-2916.1994.tb04621.x

[43] Ibach H, Lüth H. Solid-State Physics - an Introduction to Principles of Materials Science. Berlin: Springer; 2009. Chap. 11

[44] Grosso G, Parravicini GP. Solid State Physics. 2nd ed. Amsterdam: Elsevier; 2014. Chap. 9

[45] Ashcroft NW, Mermin ND. Solid State Physics. Andover: Cengate learning Ltd.; 1976. Chap. 27

[46] Wang S. Solid State Electronics. New York: McGraw-Hill; 1966. Chap. 7

[47] Grosse P. Freie Elektronen in Festkoerpern. Berlin: Springer; 1979. Japanese translation by A. Kinbara and M. Mizuhashi, Dennshibussei no kiso. Tokyo: Ohmsha; 1993. Chap. 13. DOI: 10.1007/978-3-642-95344-6

[48] Wada O, Ramachari D, Yang C-S, Pan C-L. Interrelationship among dielectric constant, energy band parameters and ionicity in multi-component oxide glasses revealed by optical- and THz-band spectroscopy. Journal of Non-Crystalline Solids. 2021; **573**:121135/1–10. DOI: 10.1016/j.jnoncrysol.2021.121135

Section 2

2D Materials for Terahertz Technology

Chapter 4

Two-Dimensional Materials for Terahertz Emission

Abdullah Alharbi, Naif Alshamrani, Hadba Hussain, Mohammed Alhamdan and Salman Alfihed

Abstract

The demand for ultrahigh-speed, lightweight, low-cost, and defect-tolerant electronic devices drives the industry to switch to terahertz (THz) technologies. The use of two-dimensional (2D) materials has massively increased in THz applications due to their appealing electronic and optoelectronic properties, including tunable bandgap, high carrier mobility, wideband optical absorption, and relatively short carrier lifetime. Several 2D-material-based emitters, modulators, and detectors have been fabricated and examined. In this context, considerable research has been going on for 2D-material-based THz emitting sources, including materials and device structure to understand the electronics and optoelectronics mechanisms occurring in the THz region. This chapter focuses on the 2D-material-based emitters with insights into the background, the physical principle of photoconductive THz emitters, the 2D materials' properties, and the research trends in the fabrication and characterization of the THz sources based upon 2D materials.

Keywords: 2D materials, terahertz devices, photoconductive THz emitter, 2D-material-based emitters, THz emission

1. Introduction

When looking back into the IEEE Transactions on Microwave Theory and Techniques, it is noted that "the first occurrence of the term *terahertz* (THz) in this Transactions is attributed to Fleming [1] in 1974, where the term was used to describe the spectral line frequency coverage of a Michelson interferometer" (page 1, [2]). Nowadays, THz radiation refers to those frequencies/wavelengths spanning 0.1–10 THz range (30–3000 μm) of the electromagnetic spectrum [3] (see **Figure 1**). However, due to generation and detection obstacles, the THz range remains one of the least discovered spectra [2]. For example, these broadband waves, at relatively high frequencies, cannot be emitted through conventional optical and electronic technologies due to the limitations on the wavelength band and the energy gap, respectively [3]. Nevertheless, the advancements in these technologies have been utilized toward assisting with the THz radiation by accelerating electrons and employing nonlinear mediums [4].

IntechOpen

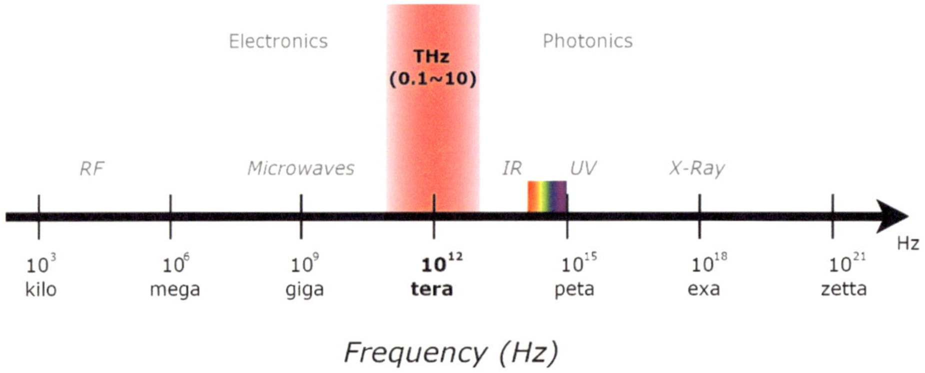

Figure 1.
Illustration of the electromagnetic spectrum in which the terahertz radiation lies in between the infrared and microwave radiations.

Terahertz technology's interest and application have grown over the past years [5]. The number of articles covering the application aspect of such a technology in different fields has exceeded that of the challenges presented in THz technology (i.e., obstacles within THz generation and detection) [6]. This technology has found itself in various scientific disciplines, from physics to biology and medical sciences [7]. The technological applications are just as vast, including outdoor/indoor communications, drug detection, security screening, semiconductor wafer inspection, semiconductor large-scale integrated (LSI) inspection, agriculture, air pollution, DNA, biometrics, medicine, and food [5]. Furthermore, this technology has continually evolved over the years as new materials and operational mechanisms are being used for the detectors and sources [8–10].

The THz sources can be classified under three categories: broadband sources (pulsed techniques), narrowband sources (continuous-wave techniques), or incoherent thermal sources [8]. Under these three categories, various sources of radiation have been investigated. Examples of such sources are mercury, silicon carbide (SiC) globar, cosmic background, quantum cascade lasers, photomixing devices (i.e., uni-traveling-carrier photodiode), free electron lasers, and backward-wave tubes or carcinotrons [5, 8, 11]. More importantly, when it comes to fundamental science/research, free electron lasers often stand out as they offer an improved signal-to-noise ratio in addition to emitting high-powered terahertz light (i.e., 10 W) for specific frequency ranges. Unfortunately, specific applications have strict limitations on the size, weight, and cost; hence such a source cannot be tolerated, and from there, researchers and scientists focused on investigating and improving other alternatives [5, 8, 11].

When looking into broadband THz emitters, it is noted that optical rectification and photoconduction device mechanisms are often employed [8]. The photoconductive emitter is an optoelectronic device that consists of photoconductive materials, photoconductive electrodes, and a lens to steer the emitted beam [3, 12]. While the THz emission by optical rectification approach can be realized by using an electro-optic (EO) medium and a nonlinear mixing, the use of ultrafast laser determines the duration of the quasi-static nonlinear response and thus the envelope of the emitted THz plus [13]. In the end, the photoconductive schemes can only be characterized by the breakdown voltage, carrier mobility, carrier lifetime, and dark resistivity [3]. Furthermore, over the years, scientists and researchers have investigated different materials for photoconductive schemes such as indium aluminum arsenide (InAlAs), indium gallium arsenide (InGaAs), gallium arsenide (GaAs), and a combination of group III-VI materials [3]. Those examples were some of the most investigated materials for photoconductive devices. Nevertheless, progress in nanofabrication technologies has enabled the discovery of other alternative materials (i.e., two-dimensional (2D) materials) that could offer optimum photoconductive characteristics [14, 15].

Two-dimensional materials have shown massive progress over the last decade in which their compelling/appealing electronic and optoelectronic characteristics make them a potential alternative for THz applications. Such characteristics include high carrier mobility, low carrier lifetime, tunable bandgap, ultra-broadband, optical absorption, and response [14, 15]. Its electrical and optical characteristics could be altered by changing their physical and chemical properties by introducing dopants (i.e., Molybdenum disulfide), strain, or an electric field [14, 15]. Several kinds of 2D materials have been explored over the years, such as graphene, transition metal dichalcogenides (TMDs), black phosphorus, and MXenes, in addition to other prospective candidates (i.e., germanium sulphide (GeS), hexagonal boron nitride (hBN), organic-inorganic hybrid perovskites, etc.) [15]. More importantly, several kinds of 2D-material-based emitters have been fabricated and examined, as seen in the literature [16–20].

This chapter will focus on 2D-material-based THz emitters with insights into the background, the physical principle and principal components of THz emitters, and the research's trends in the fabrication and characterization of the 2D materials devices structures. In Section 2, different structure designs of photoconductive THz emitters are shown and discussed. In Section 3, the 2D material properties are reviewed. This review includes its mechanical, structural, electrical, and optical priorities. Ultimately, Section 4 presents the recent advances and development in layered 2D materials and heterostructures for THz emitters.

2. Photoconductive terahertz emitter

The THz emission by a photoconductive emitter can be performed using ultrafast laser pulses and an applied bias field, E_b, taking into consideration the bandgap and the breakdown voltage of the material. In general, The THz emission mechanism will start under the optical excitation by ultrafast (femtosecond) laser pulses when it is biased at a bias field, E_b. When the optical pulses of the laser (with a bandgap higher than the band gap of the photoconductive material) hit the photoconductive gap, the free carrier will be generated and the bias field, E_b, drives the device to emit electromagnetic radiation at the THz frequencies [3]. The emission characteristics of THz devices primarily depend on the characteristics of the photoconductive material used and the design of the device. One of the promising materials is the 2D material, which will be discussed in detail in Sections 3 and 4 of this chapter. First, the photoconductive designs are discussed in the following paragraphs. The photoconductive THz emitter design can be classified into microstructure and nano(plasmonic) structures.

In the literature concerning microstructure photoconductive emitters, several studies have presented the influence of the design on the emitter performance. Madéo *et al.* adjusted electrode spacing of the interdigitated photoconductive antennas for tunning frequency of THz. They used THz-time domain spectroscopy (TDS) to measure the pulsed electric field and emission spectra. They observed that decreasing the electrode gap from 20 to 2 μm has shifted the emitted spectra peak from 0.73 to 1.33 THz, and this demonstrated that faster space-charge screening of the bias field occurs as the electrode spacing is reduced [21]. In addition, our previous work on the design and fabrication of photoconductive THz emitter based upon SI-GaAs and SI-InP investigated the influence of the bow-tie antenna structure on the THz emission performance and found that there is a correlation between the structure sharpness and the measured bandwidth of the photoconductive THz emitters [9].

In the nano (plasmonic) structures photoconductive emitter, Berry *et al*. presented the plasmonic photoconductive THz antenna for the first time, as shown in **Figure 2**. They related the new concept of the photoconductive antenna with the high quantum efficiency, which led to a significant increase in the optical-to-terahertz conversion efficiency [22].

Singh *et al*. fabricated an interdigitated photoconductive antenna, a THz emitter that eliminates the need for micro-lens array focusing. This avoids the saturation effect at a higher optical excitation density since photoexcitation can be done in larger areas. However, the device was fabricated as a microstructure and compared with the plasmonic structure device. The plasmonic photoconductive emitter emitted THz radiation power twice as much at 200 mW optical excitation [23].

Yardimci *et al.* reported a high THz radiation power with a notable improvement in optical-to-terahertz conversion efficiency [24]. They presented a novel design for a photoconductive THz emitter based on plasmonic (nano-structured) electrode. They enhanced optical-to-terahertz conversion efficiencies by a significant margin by incorporating plasmonic (nano-structured) electrodes within the

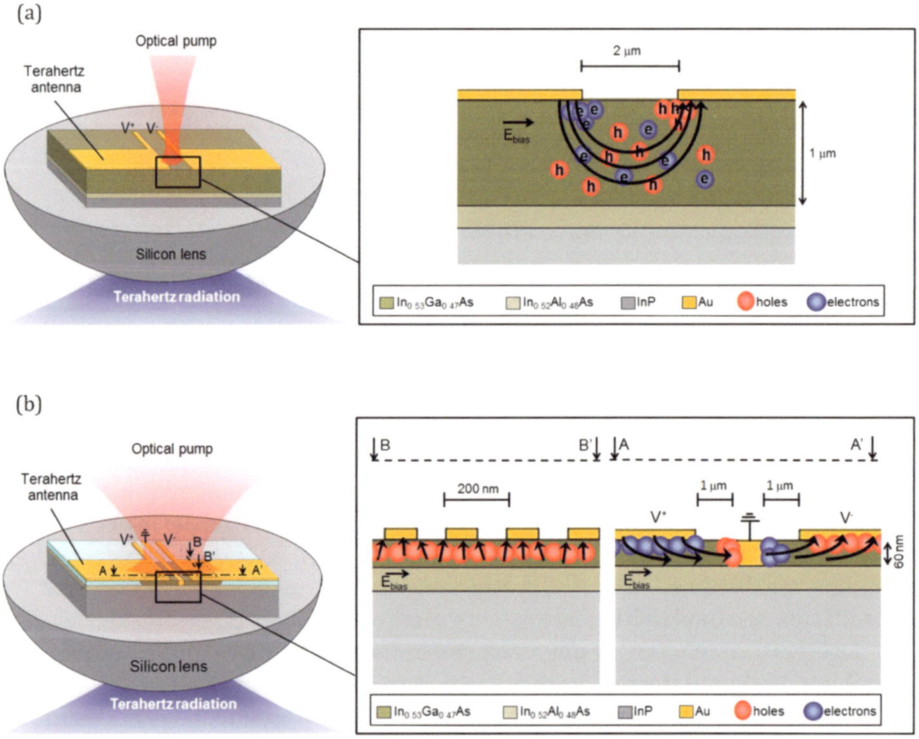

Figure 2.
(a) A microstructure photoconductive terahertz emitter with a photoconductive gap of 2 μm and (b) a nanostructure (plasmonic) photoconductive terahertz emitter based on nanoscale electrode with a spacing of 100-nm between the electrodes. *© IOP publishing. Reproduced with permission. All rights reserved [22].*

photoconductive gap, this increased time-varying dipole moments. As a result, about 1.6% optical-to-terahertz conversion efficiency was demonstrated, as the radiation power levels were up to 3.8 mW at an optical pump power of 240 mW over the 0.1–5-THz frequency range.

In addition, Berry et al. studied and fabricated the THz emitters within a logarithmic spiral antenna. The emitters were formed by an array of plasmonic photoconductive emitters, that were 3 by 3. Moreover, when optically pumped by 320 mW the device emitted 1.9 mW of pulsed THz radiation over the 0.1–2 THz frequency range. This design, as shown in **Figure 3**, offered much greater power THz pulses compared to other plasmonic photoconductive THz emitters [25].

The saw-toothed plasmonic contact electrode's structure has been studied by Zhang *et al*. A power enhancement is observed using a saw-toothed plasmonic logarithmic spiral photoconductive antenna (more than 10 times) [26]. The enhancement is accosted with the strong electric field obtained by the novel structure, as shown in **Figure 4**.

Chen *et al*. conducted an experiment on GeSn plasmonic THz photoconductive antenna [27]. The experiment compared a bowtie antenna with an active area of 20 × 20 μm^2 and a gap size of 10 μm and plasmonic gratings with a spacing and pitch size of 230 nm and 480 nm, respectively. The resulting bandwidth was up to 2 THz and SNR was 50 dB. The GeSn plasmonic photoconductive antennas prototype was fabricated through a metal-oxide-semiconductor.

In addition, based on surface plasmonic resonance (SPR), Zhiguang *et al*. [28] proposed a nano cylinder array photoconductive antenna structure by analyzing the interaction between incident light and substrate. It showed an 87% optical transmission into the GaAs substrate, which was 5.8 times the optical transmission of conventional photoconductive antennas.

The field of design and fabrication of photoconductive THz emitters is massively researched using III-V semiconductors. However, using of 2D materials for these emitters is a promising field for study and research. Therefore, the upcoming section will discuss the distinct properties of the 2D materials.

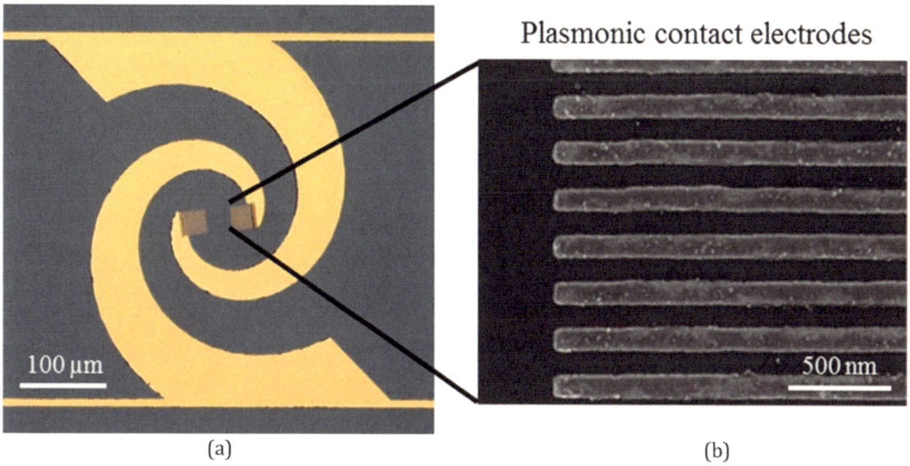

Figure 3.
The microscope (a) and SEM (b) images of the fabricated sample on an LT-GaAs substrate showing: (a) the integration/combination of the plasmonic metal electrodes within the logarithmic spiral antenna and (b) the metal electrodes. This figure is reprinted from ref. [25] with the permission of AIP publishing.

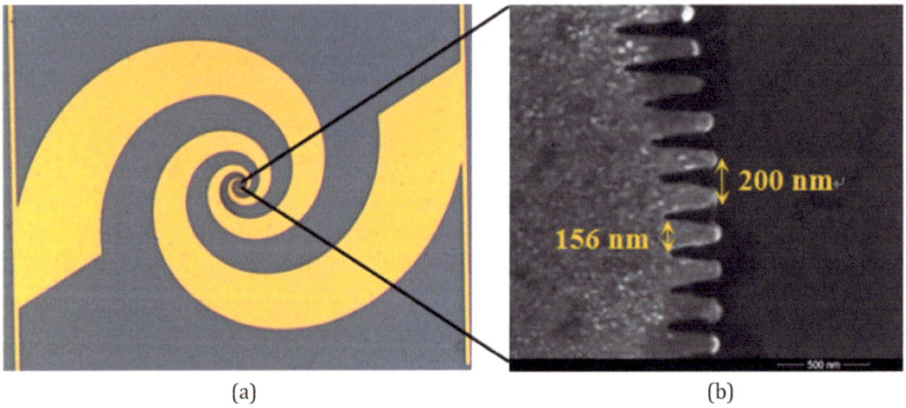

Figure 4.
The microscope (a) and SEM (b) images of the fabricated sample show: (a) the integration/combination of the saw-toothed plasmonic metal electrodes with the logarithmic spiral antenna and (b) the saw-toothed electrodes [26].

3. Properties of 2D materials

Two-dimensional materials have recently fascinated researchers' attention due to their distinct properties. Research over the last decade has made huge progress in obtaining high-performance 2D materials. This section discusses the mechanical, structural, electrical, and optical priorities of 2D materials.

3.1 Mechanical properties of 2D materials

The mechanical properties of 2D materials are paramount to uncovering aspects of the material's behavior and how the material will act under external forces. Hence, it is vital to study the mechanical properties needed to discover the strong points in 2D materials and build novel applications and devices. The mechanical properties can be studied theoretically by the density functional theory (DFT) and experimentally. For example, graphene is a two-dimensional material. Graphene is the most robust material ever, and its bandgap energy is zero eV, so the existence of graphene in the semiconductor industry is limited. However, the TMDs, such as molybdenum disulfide (MoS_2), tungsten disulfide (WS_2), molybdenum diselenide ($MoSe_2$), and tungsten diselenide (WSe_2), are potential alternative material of graphene due to the presence of their bandgaps. As well in TMDs, bandgaps can convert from an indirect bandgap in bulk into a direct bandgap in monolayer.

The properties of 2D materials have high anisotropy between the in-plane and out-of-plane mechanical properties. The close atoms are linked together via covalent bounds in the same atomic plane leading to low defects density and strong in-plane mechanical properties. However, the out-of-plane are the interlayers that are stocked together by weak van der Waals interactions. The layers can slide easily when shear stress is employed [29]. Therefore, aside from graphene, the mechanical properties of MoS_2 also have been studied in-plane. The measured in-plane stiffness and effective Young's modulus of monolayer MoS_2 are about 182 ± 14 N.m-1 and 280 ± 21 GPa, respectively, which is less than graphene's stiffness [30]. In addition, Cooper *et al.* [31] measured the stiffness theoretically depending on DFT and found that 2D Young's modulus of monolayer MoS_2 is 130 N/m and intrinsic strength is 16.5 N/m. As a result, it is unlocked for a broad of electronics and optoelectronics applications of monolayer TMDs such as fixable sensors and the ability to tune the bandgap by applying external strain.

3.2 Structural and electronic properties of 2D materials

Two-dimensional materials can be found in various forms, such as an element (graphene), metallic compounds (MoS_2, $MoSe_2$, WS_2, and others), nonmetallic compounds (hexagonal boron nitride (h-BN)), organics (polymer), and salt (clays) [32]. The shapes of the material are bulk, monolayer, bilayer, or multilayer. The crystal structure of TMDs is layered, and each layer consists of a hexagonal transition metal atoms sandwiched between two planes of hexagonally arranged dichalcogen atoms. The metal and chalcogens are covalently bonded within a layer stuck together by van der Waals bonds [33]. Thus, all TMDs have a hexagonal structure [34]. The atomic arrangement in the structure affects the electronic structure, vibrational modes, and optical properties. The atomic arrangement was designed with a particular phase 2H semiconductor phase, and metallic 1 T phase.

The 2D layered materials stacks are very different from the conventional bulk 3D heterostructures, as each layer acts simultaneously as the bulk material and the interface, reducing the charge displacement inside each layer. Nevertheless, the charge transfers between the 2D layers are massive. Thus, the band structure could be engineered by inducing large electric fields. The band structure can be engineered via alignment between the nearby crystals, number of layers, the type of the materials' surface recombination, and charge transfer [35].

3.3 Optical properties of 2D materials

Engineering of the 2D materials' optical properties can be achieved by changing one (or more) of the parameters, including material thickness and material type/structure, or by applying external strain [36]. The bandgap of 2D materials covers a wide range: from zero band gap (semimetal such as graphene) to the insulator (wide band gap). Thus, 2D materials are very attractive for optoelectronics applications, such as emitters, detectors, and optical modulators [37]. Additionally, our earlier work has established security applications as a promising field of optoelectronics applications of 2D materials [38]. The bandgaps of 2D materials have been subject to enormous studies experimentally (e.g., through photoluminescence) or theoretically through (DFT) modeling. Ketolainen *et al.* theoretically calculated the optical gap using the TD-HSE06 approach for a wide gap range of 1–6 eV of various 2D materials. The values agreed with the experimental optical gaps and the GW + BSE calculations. **Figure 5** shows that the theoretical absorbance depends on the TD-HSE06 method, HSE06 functional, and PBE functional for P_4, WSe_2, MoS_2, GaP, GaN, AlN, CF, and BN [39].

3.4 Synthesis and fabrication of 2D materials

Since the successful isolation of graphene from graphite in 2004, enormous efforts have been made to synthetize and fabricate 2D materials [40]. Typically, synthesis methods are classified as top-down or bottom-up. The top-down method involves isolation of a 2D layered sheet from bulk crystal using reduction methods, such as mechanical cleavage, liquid vapor exfoliation, or electrochemical reduction.

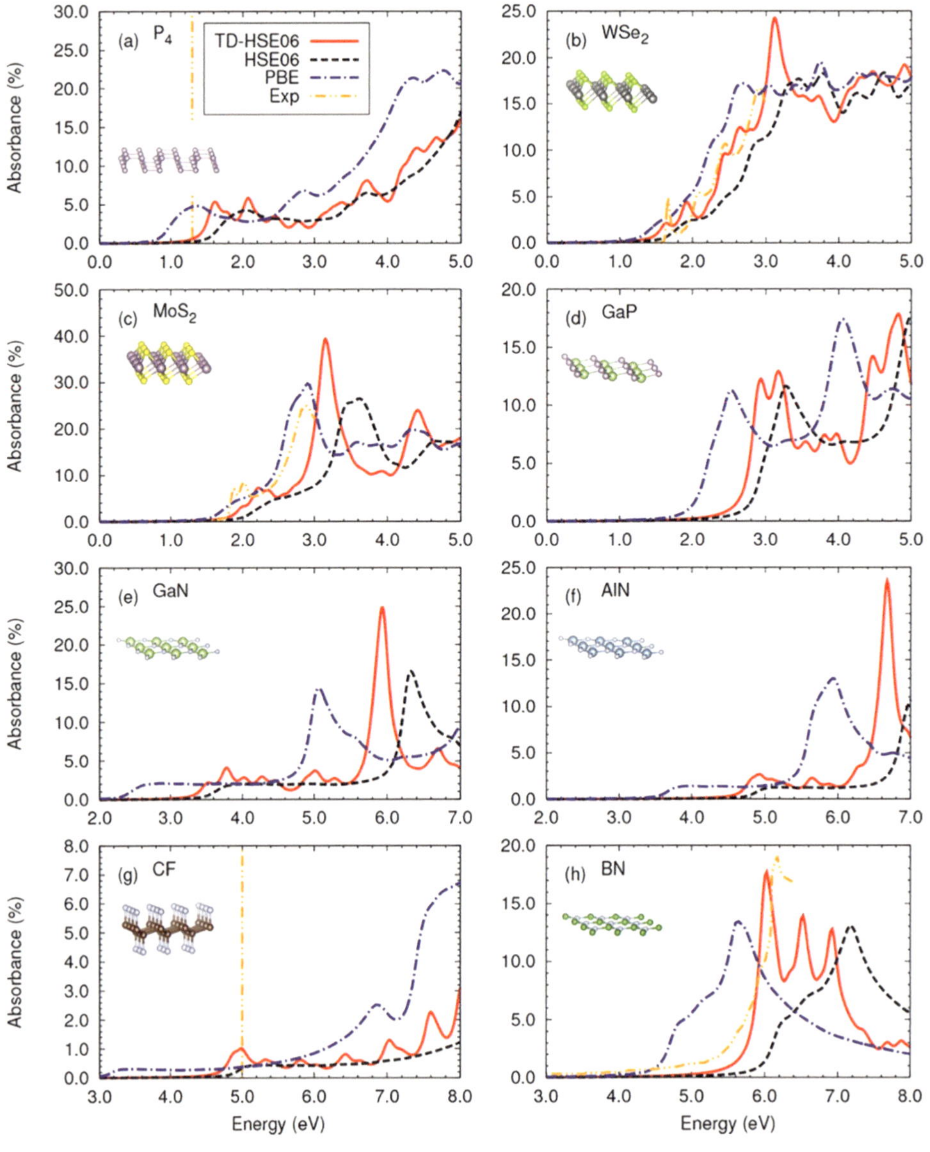

Figure 5.
This illustrates the in-plane optical absorbance using the TD-HSE06 method, HSE06 functional, and PBE functional for 2D layers: (a-h) P_4, WSe_2, MoS_2, GaP, GaN, AlN, CF, and BN, respectively. Moreover, the figure shows the experimental optical absorbance curves of WSe_2, MoS_2, and BN and the first excitation peaks in the experiments (orange curve). The atomic structures of the 2D materials are illustrated in the insets panels, "reprinted (adapted) with permission from ref. [39]. Copyright {2017} American Chemical Society."

In the bottom-up method, 2D material layers are produced from atomic or molecular precursors. This method typically involves the use of advanced techniques, such as molecular epitaxial growth (MBE), chemical vapor deposition (CVD), physical vapor deposition (PVD), and pulsed laser deposition (PLD). A representative schematic of 2D materials synthesis is shown in **Figure 6** [32].

Each bulk layer of 2D material consists of dangling bond-free layers that are weakly bound to each other via van der Waals force. By using the mechanical exfoliation method, a single layer or a few layers can be isolated from bulk materials. The applied force and adhesion to the exfoliation substrate should be controlled and overcome the van der Waals force during the exfoliation process. Typically, the mechanical exfoliation method produces the highest quality, higher crystallinity with fewer defect densities, and less impurity contamination [41]. However, the yield of this approach is very low and the method is time-consuming because of the random nature of flake thickness, size, number of flakes, and surface morphology is uncontrollable. Liquid exfoliation and chemical exfoliation techniques can be used to overcome the low yield. However, the exfoliated materials produced through these techniques are of relatively low quality and have small flake sizes. To improve the yield, we coated SiO_2 substrate with a sub-5 nm polymeric poly (vinyl alcohol) film thickness prior to exfoliation in our previous study [42]. The numbers and sizes of the produced flakes are much larger than those obtained using the conventional exfoliation method while maintaining the quality of the produced film.

As the exfoliation approach is limited in terms of large-scale production, various material growth methods have been studied to produce large-scale 2D materials. Of these, the CVD method is the most common because it allows large-scale and good-quality 2D materials to be produced. In addition, it is a low-cost method compared to MBE, PVD, and PLD. Typically, CVD is a high-temperature process used to synthesize various 2D materials on the substrate from solid, gas, or liquid precursor in a tube furnace. Typically, graphene is grown on metal substrate (e.g., Cu or Ni) using mixtures of gases. The transfer of large-area graphene films to a fabrication substrate, such as SiO_2, may lead to degradation of the electrical optical properties of the film. Alharbi et al. described high-quality, large-sized monolayer WS_2 grown on SiO_2 substrate using tungsten oxide (WO_3) and sulfur (S) solid precursors [43]. Others reported wafer-scale MoS_2 grown using metal–organic CVD [44]. Although these growth methods are scalable and produce large-size films, the quality of the films

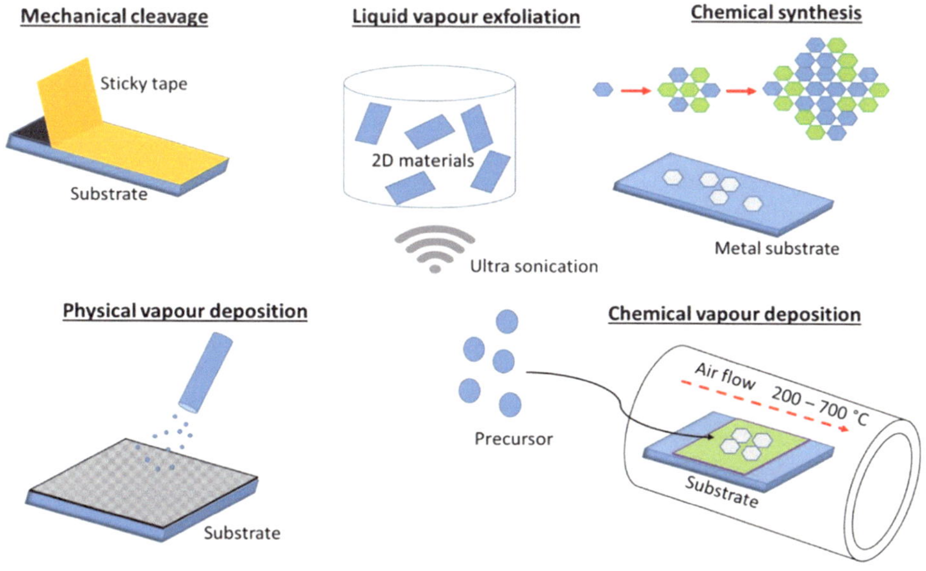

Figure 6.
A schematic of various 2D synthesis processes [32].

is lower than that of films produced using mechanical exfoliation. For example, the produced materials using CVD method are polycrystals and suffer from grain boundary and structural defects. Some of these defects may not be an issue in terms of THZ applications, the defect that suppresses the carrier lifetime.

The direct growth of heterostructure is still a major challenge. The produced heterostructures suffer from problems, such as uncontrollable growth directions, low-quality materials, and limited material combinations. Thus, most 2D material-based heterostructures are built using layer transfer methods. Suk et al. described the transfer process of CVD-grown monolayer graphene onto other substrates [45]. A deterministic dry transfer method for 2D materials that allows material positioning to be controlled has also been developed [46]. This method led to explorations of novel applications and new phenomena in 2D materials [46]. Moreover, a gold-assisted layer transfer process has been developed to overcome the adhesion of CVD-grown MoS_2 to SiO_2 and enable the controllable transfer of the film [47]. Although layer transfer methods enable many novel heterostructures to be built, the sizes of these structures remain limited.

From here, it is clear that 2D materials could become a promising alternative for THz emitters. Therefore, the upcoming section will discuss the integration and development of such materials for THz emission applications.

4. The 2D material-based THz emitters

The THz emitter is a key enabler of many THz systems. However, a wide variety of alternative techniques have been developed for THz emissions, such as nonlinear electro-optic crystal [48] and air plasma [49], to obtain THz waves of higher intensity and especially broader bandwidth. Nevertheless, these sources do not meet the current application demands. Thus, owing to their extraordinary electro-optical properties, the 2D materials-based emitter is trending in THz technologies and is expected to solve this issue. This section will focus on 2D-material-based THz emitters, including layered 2D materials and heterostructures.

4.1 Layered 2D materials

Two-dimensional materials such as graphene and MoS_2 have very high carrier mobility, surface conductivity, and a very short carrier relaxation time compared to traditional semiconductors [50]. These unique advantages can open up many new research ideas and dramatically improve device performance. Thus, 2D materials have attracted enormous research interest for THz applications. The following subsections discuss THz emitters based on layered 2D materials.

Graphene has attracted researchers' attention since its first successful separation from graphite in 2004 due to its extraordinary proprieties. Several studies have been carried out on graphene for THz applications. Graphene-based THz emitters were introduced early in 2007, where optically or electrically pumped graphene was expected [51, 52]. Then, a pumped THz laser with a Fabri-Perot resonant cavity design based on an optically pumped graphene-heterostructure was proposed in 2009 [53]. **Figure 7** illustrates the structure of the device where the optical excitation with the energy of $\hbar\Omega$ generates electrons and hole pairs in the graphene and subsequently acquires population inversion to generate THz radiation.

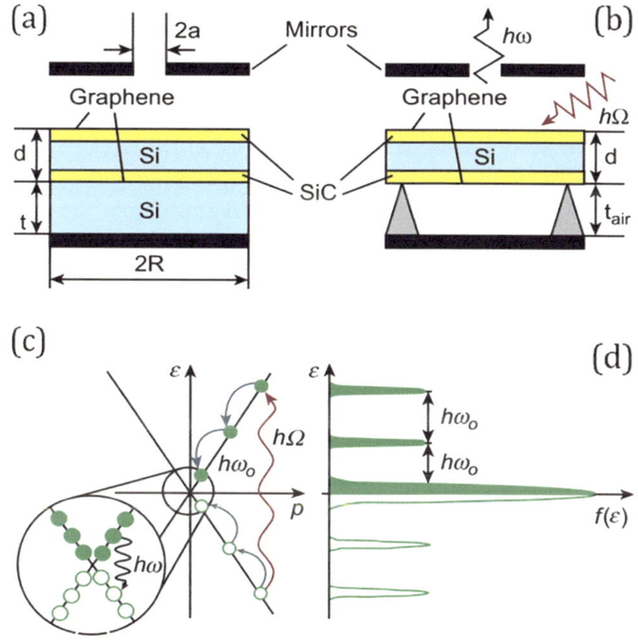

Figure 7.
Schematic view of the laser structures with (a) Si separation layer (b) air separation layer, (c) laser pumping scheme (d) electron and hole distribution functions. Figures were reprinted with permission from ref. [53].

With this in mind, a graphene-based THz laser based on optically pumped multiple-graphene-layer structures separated with a metal slot-line or a dielectric waveguide was proposed [54]. The THz lasing based on graphene P-i-n was demonstrated under certain experimental conditions [16]. Additionally, graphene-based plasmonic oscillator THz generators were proposed. Furthermore, it was found that tuning the intrinsic plasmons of graphene can generate radiation up to 1 THz [16].

Moreover, the current injection graphene-based THz emitter has been demonstrated to achieve higher power. Tong *et al*. demonstrated THz sources (1–3 THz) created with graphene field effect transistors, where the graphene has been attached to a double-patch antenna and an on-chip silicon lens [55]. The fabricated THz sources operate at room temperature and produce significantly higher power than conventional THz sources. The graphene-antenna device was fabricated on a high-resistivity silicon chip covered with a 300 nm-thick silicon oxide layer. Then, they were bonded back-to-back on an elliptical silicon lens. The designed antenna has a size of 45 × 31 μm. Graphene film was placed between the two metal patches of palladium or gold. This double patch also served as the source and drain contacts for the graphene transistor. According to their simulation results using a high-frequency structure simulator (HFSS), the impedance spectrum indicates an optimal operating frequency of 2.1THz. Therefore, the device can be used as a THz emitter or detector.

Coherent THz emission from monolayer graphene has been reported by Maysonnave *et al*. [56]. They experimentally demonstrated the process of exciting the graphene by femtosecond optical pulses to generate THz emitting in the range from 0.1 to 4 THz. The emission occurred because of a second-order nonlinear effect, which is dynamical photon drag that is induced by the transfer of light momentum to carry through ponderomotive electric and magnetic forces [57].

Recently, Zhang *et al*. demonstrated a new hybrid photoconductive to overcome the fundamental trade-off between picosecond ultrafast pulses power output. The photoconductive switch was designed by optimizing a hot-carrier fast lane using graphene on silicon. Because of the combination of the strong absorption in bulk semiconductors with a graphene layer as a hot carrier fast lane, the proposed devices emitted a high-power THz field, with up to 80 times amplitude improvement compared to the devices without graphene [58].

4.1.1 Transitional metal dichalcogenides (TMDS)

Layered transitional metal dichalcogenides such as MoS_2 and WS_2 are a kind of 2D material that have remarkable optical properties. Unlike gapless graphene, TMDs have tunable band gaps that can be changed from indirect to direct band gaps with a number of layers.

Molybdenum disulfide (MoS_2) is the most studied TMDs material owing to its striking properties for optoelectronic applications. As the demand for the working frequency of devices increases toward the THz region for high-speed applications, the unique advantages of MoS_2 are expected to play a major role. For instance, the photoconductivity response time can reach 350 fs from CVD-grown monolayer MoS_2 and 1 Ps from trilayer MoS_2 and monolayer WSe_2 [59]. Huang *et al*. studied the THz pulse emission from layered MoS_2 film. They observed that MoS_2 can generate from 0.1 to 3.5 THz radiation time-domain surface emission spectroscopy under the excitation of linearly polarized femtosecond laser with the reflection configuration, as shown in **Figure 8**(a). The radiation amplitude has a linear dependence on the incident pump influence, which can confirm the second-order nonlinear process. **Figure 8(b)** reveals that the single-cycle THz pulses produced from the three samples (MoS_2, graphite, and InAs) have similar waveforms [60]. However, the THz radiation of MoS_2 has a longer pulse delay because of the combination of penetration depth and rugged surface. Nevertheless, the research of TMDs-based THz emitter is ongoing and most

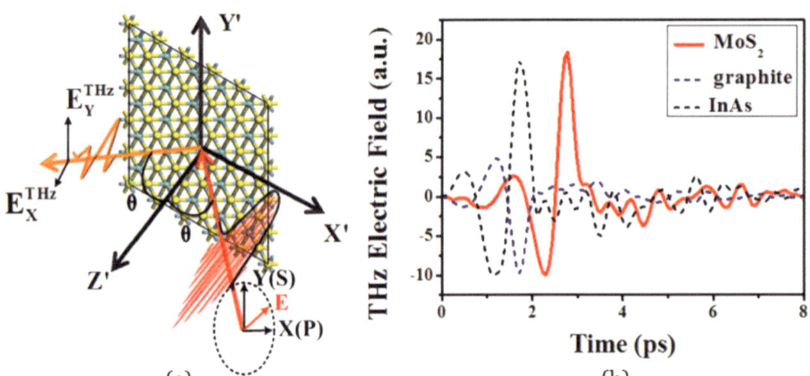

Figure 8.
(a) Schematic illustration of THz radiation in a reflection configuration. (b) THz pulses in time domain produced from MoS_2, graphite, and InAs layered crystal. "Reprinted (adapted) with permission from ref. [60]. Copyright {2017} American Chemical Society."

of the research focuses on optimizing and producing controllable various types of TMDs for THz applications.

4.1.2 Topological insulator

Topological insulators (TIs) are promising 2D materials for the next generation of optoelectronic applications. Among the TIs materials, Bi_2Se_3 has attracted tremendous research interest in the scientific community for THz emission. An efficient THz emission from the Bi_2Se_3 basal plane was observed upon femtosecond optical excitation [61]. Moreover, Giorgianni *et al*. demonstrated that Dirac surface states in Bi_2Se_3 TIs induce a strong nonlinear THz response [62]. Additionally, surface crystallography plays a major role in THz emission. For instance, Bi_2Se_3 grown on Al_2O_3 exhibits an anisotropic response with a strong modulation of the THz while in its phase, while its thin film grown on Si nanocrystals leads to THz emission in an opposite phase [63].

In another study, THz emission from Bi_2Te_3 nanofilms was studied. In their study, they used various thicknesses of Bi_2Te_3 nanofilms grown on Al_2O_3 substrates and excited the film by femtosecond laser pulses [64]. An efficient linear polarization of tunable THz radiation and high-quality chiral THz waves was demonstrated by controlling the pump laser pulse polarization, incident angle, and sample azimuthal angle.

Furthermore, 2D-TIs alloys: binary, ternary, and quaternary with a specific composition have been investigated in many studies [65, 66]. For example, Kuznetsov et al. studied the THz emission proprieties of the $Bi_{2-x}Sb_xTe_{3-y}Se_y$ (BSTS) topological insulators films grown by the MOCVD method on a sapphire substrate, with different thicknesses and chemical compositions [65]. They conclude the THz radiation intensity depends on the film thickness, and the highest THz radiation intensity observed in the island film has a total thickness of about tens nanometers with a composition close to the Ren's curve.

Overall, the TIs antenna is very proposing for THz emitters because they are low cost, extremely thin, and easy to integrate into nano-photonic devices. Also, their conversion efficiency is comparable to state-of-the-art THz emitter and could be further improved, for example, by applying plasmon gratings.

4.2 Heterostructures

Heterostructures promise to be an enabling key for novel optoelectronic devices. There are two types of heterostructures: 1) Van der Waals (vdW) heterostructures are composed of different 2D and 2) heterogeneous integration with traditional materials and devices.

Van der Waals heterostructures and devices made by stacking different 2D crystals on top of each other attract many researchers as their properties change, leading to novel hybrid properties and enhancement of electronics and optoelectronic properties [42, 67]. Many THz emitters based on van der Waals heterostructures have been proposed so far. MoS_2 monolayer sandwiched between two graphene films reveals an extraordinary enhancement in the transient THz response compared to monolayer MoS_2 and graphene [68]. The enchantment was attributed to the THz response in this heterostructure produced from the sub-bandgap below the band gap of MoS_2. Moreover, THz pulses with a bandwidth of 2.5 THz have been reported based on $MoSe_2$ and ReS_2 vdW heterostructure [69]. In this study, they used different stacking orders and measurement techniques to address the photocarrier dynamics. The emitted THz pulse is reversed by the interchange of the stack ordered, indicating that the

THz radiations originate from the plane current. The corresponding frequent spectrum of the waveforms has a peak at 1.0 THz and a bandwidth of 2.5 THz. Other vdW heterostructures such as WS_2/hBN have also been studied for THz photoconductivity [70]. Far infrared and THz emitting diode also has been proposed based on graphene/black phosphorus (BP) and graphene/MoS_2 by Ryzhii *et al*. [71].

One of the advantages of 2D materials is that they are easy to integrate with traditional optoelectronic materials, systems, and devices. These heterogeneous integrations enable many novel phenomena and applications, particularly in the THz regime. Typically, the incorporation of 2D materials into other structures is accomplished by either direct growth or thin-film transfer methods. Graphene transferred to SiO_2/Si substrate was used to study the THz emission spectroscopy on the graphene-based interface and evaluate the interface build-in potential and trapped charge dynamics [72]. In fact, various heterogeneous integrations for many THz applications including MoS_2/Si and direct growth of Bi_2Se_3 onto GaAs substrate [50, 73].

The most advanced type of heterostructure-based THz emitter is the spintronic emitter, which offers many advantages, including ultra-broad bandwidth, high efficiency, and ease of control of radiation parameters through polarization, amplitude, and phase. Typically, the THz emitter devices are made of ferromagnetic and nonmagnetic (FM/NM) thin metal film heterostructures to generate the THz radiation when the heterostructures are irradiated with optical femtosecond laser pulses [74]. One of the main advantages of 2D materials is the presence of spin-orbit coupling, which is critical for spintronics THz emitters [75, 76].

An efficient spintronic THz emitter based on ferromagnet/heavy metal/topological insulator heterostructures was demonstrated. The THz radiation amplitude of the Co/Bi/Bi_2Te_3 heterostructure increases by 198% compared to the device without an intercalated Bi film, as shown in **Figure 9** [77].

Highly efficient THz emission from the Bi_2Se_3/Co heterostructures modified with ultrafast spin injection has been reported [78]. The Bi_2Se_3/Co heterostructure reveals a much larger THz pulse amplitude compared to a single film. Moreover, efficient spin current injection from Co to monolayer MoS_2 at Co/MoS_2 structures via THz emission has been demonstrated [79]. The sample was irradiated with 50 fs linearly 800 nm laser pulses and the magnetization direction of Co was controlled

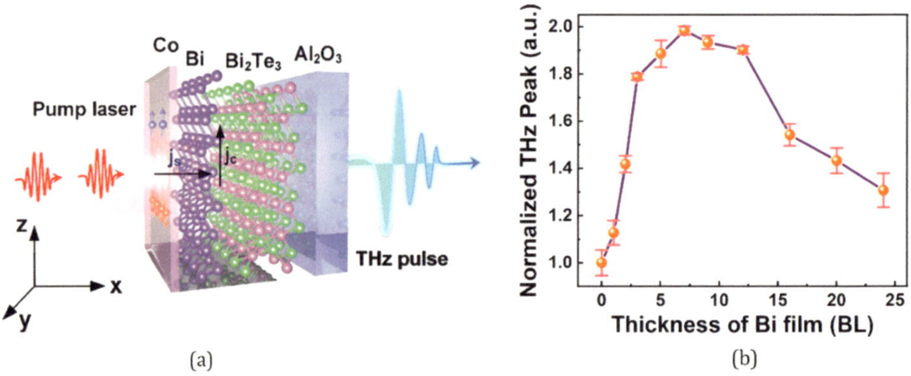

Figure 9.
(a) The schematic of the experimental measurements for THz radiation on the Co/Bi and (b) normalized THz peak amplitudes as a function of Bi film thickness. "Reprinted (adapted) with permission from ref. [77]. Copyright {2021} American Chemical Society."

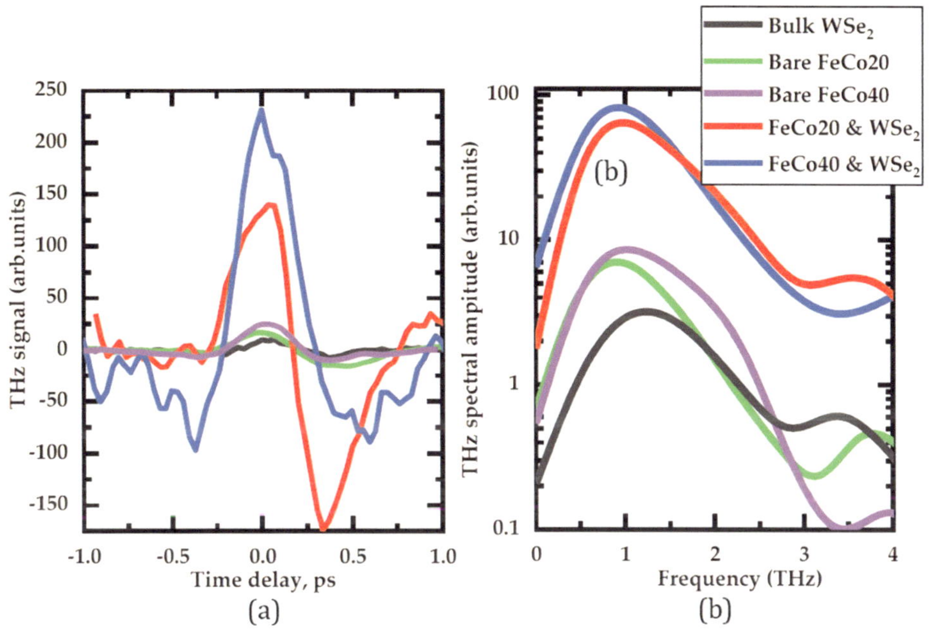

Figure 10.
(a) The THz radiation obtained by various structures in time domain and (b) the correspondent frequency spectrum [80].

by an external magnetic field. The detected THz radiation from Co/MoS_2 is strong and has an amplitude of ~2% compared to the THz emission of 1-mm thick ZnTe at the same pump fluence. Ultimately, increasing the efficiency of WSe_2/FeCo heterostructures, spintronics THz emitter has been demonstrated in reflection geometry [80]. **Figure 10** compares the THz radiation of various samples to unveil the enhancement of the THz emissions.

Overall, the 2D materials have been investigated toward high THz radiation efficiency as well as reduce the cost of the photoconductive materials. However, 2D materials are still awaiting full exploitation. Although the device fabrication is very similar to other materials-based emitters, many challenges still remain to realize efficient THz devices, such as obtaining materials with very high charge carrier mobility, low cost, and scalable. For instance, a number of research activities start on other new materials, such as gallium telluride (GaTe) [81]. Moreover, many 2D materials preparation and fabrication technologies are discussed in Section 3.4 to address many challenges related to the quality of the produced materials and scalability.

5. Conclusion

This chapter presented the 2D-material-based THz sources, particularly the photoconductive THz emitter. Several 2D materials have been employed as photoconductive materials and device structures. First, the chapter covers the basic principle of photoconductive terahertz emitter and the properties of 2D physical materials. In the end, the chapter discusses the 2D material-based THz emitters, where many forms of

2D materials and device structures are presented, including layered 2D materials and homogenous and heterogeneous heterostructures.

Conflict of interest

The authors declare no conflict of interest.

Author details

Abdullah Alharbi, Naif Alshamrani, Hadba Hussain, Mohammed Alhamdan and Salman Alfihed*
Microelectronics and Semiconductor Institute, (KACST), Riyadh, Saudi Arabia

*Address all correspondence to: salfihed@kacst.edu.sa

IntechOpen

© 2023 The Author(s). Licensee IntechOpen. This chapter is distributed under the terms of the Creative Commons Attribution License (http://creativecommons.org/licenses/by/3.0), which permits unrestricted use, distribution, and reproduction in any medium, provided the original work is properly cited.

References

[1] Fleming JW. High-resolution Submillimeter-wave Fourier-transform spectrometry of gases. IEEE Transactions on Microwave Theory and Techniques. 1974;**22**(12):1023-1025

[2] Siegel PH. Terahertz technology. IEEE Transactions on Microwave Theory and Techniques. 2002;**50**(3):910-928

[3] Alfihed S, Alharbi A. Broadband terahertz emission from photoconductive devices. In: Intelligent Electronics and Circuits - Terahertz, ITS, and beyond. London: IntechOpen; 2022

[4] Lee Y-S. Principles of Terahertz Science and Technology. New York: Springer; 2009

[5] Tonouchi M. Cutting-edge terahertz technology. Nature Photonics. 2007;**1**:97-105

[6] Dragoman D, Dragomanb M. Terahertz fields and applications. Progress in Quantum Electronics. 2004;**28**(1):1-66

[7] Davies AG, Linfield EH, Johnston MB. The development of terahertz sources and their applications. Physics in Medicine & Biology. 2002;**47**:3679-3689

[8] Ferguson B, Zhang X-C. Materials for terahertz science and technology. Nature Materials. 2002;**1**:26-33

[9] Alfihed S, Foulds IG, Holzman JF. Characteristics of Bow-tie antenna structures for semi-insulating GaAs and InP photoconductive terahertz emitters. Sensors. 2021;**21**(9):3131

[10] Bertulis K, Krotkus A, Aleksejenko G, Pačebutas V, Adomavičius R, Molis G. GaBiAs: A material for optoelectronic terahertz devices. Applied Physics Letters. 2006;**88**:201112

[11] Dhillon SS, Vitiello M, Linfield E, Davies A, Hoffmann M, Booske J, et al. The 2017 terahertz science and technology roadmap. Journal of Physics D: Applied Physics. 2017;**50**(4):043001

[12] Bacon DR, Madéo J, Dani KM. Photoconductive emitters for pulsed terahertz generation. Journal of Optics. 2021;**23**:064001

[13] Rice A, Jin Y, Ma XF, Zhang X. Terahertz optical rectification from⟨110⟩ zinc-blende crystals. Applied Physics Letters. 1998;**64**(11):1324-1326

[14] Goel N, Kumar M. 2D materials for terahertz application. Nano Express. 2021;**2**(3):031001

[15] Shi Z, Zhang H, Khan K, Cao R, Zhang Y, Ma C, et al. Two-dimensional materials toward terahertz optoelectronic device applications. Journal of Photochemistry and Photobiology C: Photochemistry Reviews. 2022;**51**:100473

[16] Basu T, Banerjee A, Vajandar S. 2D materials as THz generators, detectors, and modulators: Potential candidates for biomedical applications materials to devices. In: Terahertz Biomedical and Healthcare Technologies. Amsterdam, Netherlands: Elsevier; 2020. pp. 75-87

[17] Yao Z, Huang Y, Zhu L, Obraztsov PA, Du W, Zhanga L, et al. Interfacial THz generation from graphene/Si mixed-dimensional van der Waals heterostructure. Nanoscale. 2019;**11**(35):16614-16620

[18] Kindness SJ, Almond NW, Michailow W, Wei B, Jakob LA,

Delfanazari K, et al. Graphene-integrated metamaterial device for all-electrical polarization control of terahertz quantum Cascade lasers. ACS Photonics. 2019;**6**(6):1547-1555

[19] Xing X, Zhao L, Zhang Z, Lin X, Yu Y, Jin Z, et al. The modulation of terahertz photoconductivity in CVD grown n-doped monolayer MoS2 with gas adsorption. Journal of Physics: Condensed Matter. 2019;**31**(24):245001

[20] Fang Z, Wang H, Wu X, Shan S, Wang C, Zhao H, et al. Nonlinear terahertz emission in the three-dimensional topological insulator Bi2Te3 by terahertz emission spectroscopy. Applied Physics Letters. 2019;**115**(19):191102

[21] Madéo J, Jukam N, Oustinov D, Rosticher M, Rungsawang R, Tignon J, et al. Frequency tunable terahertz interdigitated photoconductive antennas. Electronics Letters. 2010;**611**(9):46

[22] Berry CW, Jarrahi M. Terahertz generation using plasmonic photoconductive gratings. New Journal of Physics. 2012;**14**:105029

[23] Abhishek S, Prabhu SS. Microlensless interdigitated photoconductive terahertz emitters. Optical Express. 2015;**1529**(2):23

[24] Yardimci NT, Yang S-H, Berry CW, Jarrahi M. High-power terahertz generation using large-area. IEEE Transactions on Terahertz Science and Technology. 2015;**5**(2):223-229

[25] Berry CW, Hashemi MR, Jarrahi M. Generation of high power pulsed terahertz radiation using a plasmonic photoconductive emitter array with logarithmic spiral antennas. Applied Physics Letters. 2014;**081122**(08):104

[26] Zhang X, Zhan F, Wei X, He W, Ruan C. "Performance enhancement of photoconductive antenna using saw-toothed Plasmonic contact electrodes." Electronics. 2021;**10**:2693

[27] Chen W-C, Yang S-H. GeSn plasmonic terahertz photoconductive antenna. In: 2020 45th International Conference on Infrared, Millimeter, and Terahertz Waves (IRMMW-THz), Buffalo, NY, USA. 2020

[28] Ao Z, Sun J, Cai H, Song G, Song J, Song Y, et al. Nanoplasmonic-gold-cylinder-array-enhanced terahertz source. Journal of Semiconductors. 2016;**37**:123002

[29] Zhang R, Cheung R. Mechanical Properties and Applications of Two-Dimensional Materials. Rijeka, Crotia: InTech; 2016

[30] Wang W, Li L, Yang C, Soler-Crespo RA, Meng Z, Li M, et al. Plasticity resulted from phase transformation for monolayer molybdenum disulfide film during nanoindentation simulations. Nanotechnology. Mar 2017;**28**(16):164005

[31] Cooper RC, Lee C, Marianetti CA, Wei X, Hone J, Kysar JW. Nonlinear elastic behavior of two-dimensional molybdenum disulfide. Physical review. B, Condensed Matter and Materials Physics. 2017;**87**:035423

[32] Shanmugam V, Mensah RA, Babu K, Gawusu S, Chanda A, Tu Y, et al. A review of the synthesis, properties, and applications of 2D materials. Particle & Particle Systems Characterization. 2022;**39**(6):2200031

[33] Choudhury TH, Zhang X, Balushi ZYA, Chubarov M, Redwing JM.

Epitaxial growth of two-dimensional layered transition metal Dichalcogenides. Annual Review of Materials Research. 2020;**50**:155-177

[34] Somvanshi D, Jit S. Transition metal dichalcogenides based two-dimensional heterostructures for optoelectronic applications. In: 2D Nanoscale Heterostructured Materials. Amsterdam, Netherlands: Elsevier; Jan 2020. pp. 125-149

[35] Novoselov KS, Mishchenko A, Carvalho A, Neto AHC. 2D materials and van der Waals heterostructures. Science (80-.). July 2016;**353**(6298): aac9439

[36] Mas-Ballesté R, Gómez-Navarro C, Gómez-Herrero J, Zamora F. 2D materials: To graphene and beyond. Nanoscale. 2011;**3**(1):20-30

[37] Wang M, Yang E-H. THz applications of 2D materials: Graphene and beyond. Nano-Structures \& Nano-Objects. 2018;**15**:101-113

[38] Alharbi A, Armstrong D, Alharbi S, Shahrjerdi D. Physically unclonable cryptographic primitives by chemical vapor deposition of layered MoS2. ACS Nano. 2017;**11**(12):12772-12779

[39] Ketolainen T, MacHáčová N, Karlický F. "Optical gaps and Excitonic properties of 2D materials by hybrid time-dependent density functional theory: Evidences for monolayers and prospects for van der Waals Heterostructures," Journal of Chemical Theory and Computation. Sep 2020;**16**(9):5876-5883

[40] Novoselov KS, Geim AK, Morozov SV, Jiang D, Zhang Y, Dubonos SV, et al. Electric field effect in atomically thin carbon films. Science. 2004;**306**(5696):666-669

[41] Liu Y, Weiss NO, Duan X, Cheng H-C, Huang Y, Duan X. Van der Waals heterostructures and devices. Nature Reviews Materials. 2016;**1**(9):1-17

[42] Huang Z, Alharbi A, Mayer W, Cuniberto E, Taniguchi T, Watanabe K, et al. Versatile construction of van der Waals heterostructures using a dual-function polymeric film. Nature Communications. 2020;**11**(1):1-10

[43] Alharbi A, Shahrjerdi D. Electronic properties of monolayer tungsten disulfide grown by chemical vapor deposition. Applied Physics Letters. 2016;**109**(19):193502

[44] Kang K, Xie S, Huang L, Han Y, Huang PY, Mak KF, et al. High-mobility three-atom-thick semiconducting films with wafer-scale homogeneity. Nature. 2015;**520**(7549):656-660

[45] Suk JW, Kitt A, Magnuson CW, Hao Y, Ahmed S, An J, et al. Transfer of CVD-grown monolayer graphene onto arbitrary substrates. ACS Nano. 2011;**5**(9):6916-6924

[46] Castellanos-Gomez A, Buscema M, Molenaar R, Singh V, Janssen L, Zant HSVD, et al. Deterministic transfer of two-dimensional materials by all-dry viscoelastic stamping. 2D Materials. 2014;**1**(1):011002

[47] Alharbi A, Huang Z, Taniguchi T, Watanabe K, Shahrjerdi D. Effect of substrate coupling on the performance and variability of monolayer MoS 2 transistors. IEEE Electron Device Letters. 2018;**40**(1):135-138

[48] Rice A, Jin Y, Ma XF, Zhang X, Bliss D, Larkin J, et al. Terahertz optical rectification from< 110> zinc-blende crystals. Applied Physics Letters. 1994;**64**(11):1324-1326

[49] Dai J, Liu J, Zhang X-C. Terahertz wave air photonics: Terahertz wave generation and detection with laser-induced gas plasma. IEEE Journal of Selected Topics in Quantum Electronics. 2010;**17**(1):183-190

[50] Zheng W, Fan F, Chen M, Chen S, Chang S-J. Optically pumped terahertz wave modulation in MoS2-Si heterostructure metasurface. AIP Advances. 2016;**6**(7):075105

[51] Ryzhii V, Ryzhii M, Otsuji T. Negative dynamic conductivity of graphene with optical pumping. Journal of Applied Physics. 2007;**101**(8):083114

[52] Maxim R, Victor R. Injection and population inversion in electrically induced p--n junction in graphene with split gates. Japanese Journal of Applied Physics. 2007;**46**(3L):L151

[53] Dubinov A, Alexander VY, Aleshkin M, Ryzhii TO, Ryzhii V. Terahertz laser with optically pumped graphene layers and Fabri–Perot resonator. Applied Physics Express. 2009;**2**(9):092301

[54] Ryzhii V, Dubinov AA, Otsuji T, Mitin V, Shur MS. Terahertz lasers based on optically pumped multiple graphene structures with slot-line and dielectric waveguides. Journal of Applied Physics. 2010;**107**(5):054505

[55] Tong J, Muthee M, Chen S-Y, Yngvesson SK, Yan J. Antenna enhanced graphene THz emitter and detector. Nano Letters. 2015;**15**(8):5295-5301

[56] Maysonnave J, Huppert S, Wang F, Maero S, Berger C, Heer WD, et al. Terahertz generation by dynamical photon drag effect in graphene excited by femtosecond optical pulses. Nano Letters. 2014;**14**(10):5797-5802

[57] Sun D, Divin C, Rioux J, Sipe JE, Berger C, Heer WAD, et al. Coherent control of ballistic photocurrents in multilayer epitaxial graphene using quantum interference. Nano Letters. 2010;**10**(4):1293-1296

[58] Zhang D, Xu Z, Cheng G, Liu Z, Gutierrez AR, Zang W, et al. "strongly enhanced THz generation enabled by a graphene hot-carrier fast lane," nature. Communications. 2022;**13**(1):1-7

[59] Docherty CJ, Parkinson P, Joyce HJ, Chiu M-H, Chen C-H, Lee M-Y, et al. Ultrafast transient terahertz conductivity of monolayer MoS2 and WSe2 grown by chemical vapor deposition. ACS Nano. 2014;**8**(11):11147-11153

[60] Huang Y, Zhu L, Zhao Q, Guo Y, Ren Z, Bai J, et al. Surface optical rectification from layered MoS2 crystal by THz time-domain surface emission spectroscopy. ACS Applied Materials & Interfaces. 2017;**9**(5):4956-4965

[61] Zhu L-G, Kubera B, Mak KF, Shan J. Effect of surface states on terahertz emission from the Bi2Se3 surface. Scientific Reports. 2014;**5**:1-8

[62] Giorgianni F, Chiadroni E, Rovere A, Cestelli-Guidi M, Perucchi A, Bellaveglia M, et al. Strong nonlinear terahertz response induced by Dirac surface states in Bi2Se3 topological insulator. Nature Communications. 2016;**7**:1-6

[63] Hamh SY, Park S-H, Han J, Jeon JH, Kahng S-J, Kim S, et al. Anisotropic terahertz emission from Bi2Se3 thin films with inclined crystal planes. Nanoscale Research Letters. 2015;**10**(1):1-6

[64] Zhao H, Chen X, Ouyang C, Wang H, Kong D, Yang P, et al. Generation and manipulation of chiral terahertz waves in the three-dimensional topological

insulator Bi2Te3. Advanced Photonics. 2020;**2**(6):066003

[65] Kuznetsov KA, Kitaeva GK, Kuznetsov PI, Yakushcheva GG. Generation of terahertz radiation from the island films of topological insulator Bi2-xSbxTe3-ySey. AIP Advances. 2019;**9**(1):015310

[66] Kuznetsov KA, Safronenkov DA, Kuznetsov PI, Kitaeva GK. Terahertz photoconductive antenna based on a topological insulator nanofilm. Applied Sciences. 2021;**11**(12):5580

[67] Geim AK, Grigorieva IV. Van der Waals heterostructures. Nature. 2013;**499**(7459):419-425

[68] Kumar S, Singh A, Nivedan A, Kumar S, Yun SJ, Lee YH, et al. Sub-bandgap activated charges transfer in a graphene-MoS2-graphene heterostructure. Nano Select. 2021;**2**(10):2019-2028

[69] Yang J, Jiang S, Xie J, Jiang H, Xu S, Zhang K, et al. Identifying the intermediate free-carrier dynamics across the charge separation in monolayer MoS2/ReSe2 Heterostructures. ACS Nano. 2021;**15**(10):16760-16768

[70] Krishna MBM, Madéo J, Urquizo JP, Zhu X, Vinod S, Tiwary CS, et al. Terahertz photoconductivity and photocarrier dynamics in few-layer hBN/WS2 van der Waals heterostructure laminates. Semiconductor Science and Technology. 2018;**33**(8):084001

[71] Ryzhii V, Ryzhii M, Maltsev PP, Karasik VE, Mitin V, Shur MS, et al. Far-infrared and terahertz emitting diodes based on graphene/ black-P and graphene/MoS 2 heterostructures. Optics Express. 2020;**28**(16):24136-24151

[72] Yao Z, Zhu L, Huang Y, Zhang L, Du W, Lei Z, et al. Interface properties probed by active THz surface emission in graphene/SiO2/Si heterostructures. ACS Applied Materials \& Interfaces. 2018;**10**(41):35599-35606

[73] Liu Y, Acuna W, Zhang H, Ho DQ, Hu R, Wang Z, et al. Bi2Se3 growth on (001) GaAs substrates for terahertz integrated systems. ACS Applied Materials \& Interfaces. 2022;**14**(37):42683-42691

[74] Feng Z, Qiu H, Wang D, Zhang C, Sun S, Jin B, et al. Spintronic terahertz emitter. Journal of Applied Physics. 2021;**129**(1):010901

[75] Moore JE. The birth of topological insulators. Nature. 2010;**464**(7286):194-198

[76] Wang FK, Yang SJ, Zhai TY. 2D Bi2Se3 materials for optoelectronics. Iscience. 2021;**24**(11):103291

[77] Tong M, Hu Y, Wang Z, Zhou T, Xie X, Cheng X, et al. Enhanced terahertz radiation by efficient spin-to-charge conversion in rashba-mediated Dirac surface states. Nano Letters. 2020;**21**(1):60-67

[78] Wang X, Cheng L, Zhu D, Wu Y, Chen M, Wang Y, et al. Ultrafast spin-to-charge conversion at the surface of topological insulator thin films. Advanced Materials. 2018;**30**(52):1802356

[79] Cheng L, Wang X, Yang W, Chai J, Ming Yang MC, Yang Wu XC, et al. Far out-of-equilibrium spin populations trigger giant spin injection into atomically thin MoS2. Nature Physics. 2019;**15**(4):347-351

[80] Khusyainov D, Guskov A, Ovcharenko S, Tiercelin N,

Preobrazhensky V, Buryakov A, et al. Increasing the efficiency of a spintronic THz emitter based on WSe2/FeCo. Materials. 2021;**21**(14):6479

[81] Dong J, Gradwohl K-P, Xu Y, Wang T, Zhang B, Xiao B, et al. Terahertz emission from layered GaTe crystal due to surface lattice reorganization and in-plane noncubic mobility anisotropy. Photonics Research. 2019;7(5):518-525

Chapter 5

Graphene-Based Plasmonic Terahertz Laser Transistors

Taiichi Otsuji

Abstract

This chapter reviews recent advances in the research of graphene-based plasmonic terahertz laser transistors. Optically or electrically pumped graphene works as a gain medium in the terahertz frequency range. The author's group theoretically discovered this fact and experimentally verified the single mode terahertz emission, as well as broadband terahertz amplified spontaneous emission from fabricated graphene-channel field-effect transistor (GFET) laser chips. However, its lasing threshold temperature was low (100 K) and emission intensity was weak. To drastically improve the laser performance, the introduction of graphene Dirac plasmons (GDPs) as the gain booster is promising. The author's group found a novel way to promote the current-driven instability of the GDPs in an asymmetric dual-grating-gate GFET, demonstrating room-temperature amplification of stimulated emission of terahertz radiation with the maximal gain of 9% which is four times larger than the quantum-mechanical limit when terahertz photons directly interact with graphene electrons without excitation of the GDPs. The author also proposes the active controlling of the parity and time-reversal symmetries of the GDPs as a paradigm towards ultrafast direct gain switching in the GFET lasers. Future directions to unite the gain seed and amplifier sections in a single GFET structure will be addressed with several feasible scenarios.

Keywords: graphene, plasmon, terahertz, laser, transistor, instability, PT-symmetry

1. Introduction

Exploitation of the electromagnetic resources in the terahertz (THz) spectra is the fundamental need to envision a real-world future smart society based on innovative information and communication technology (ICT) which will be led by the next generation 6G and 7G wireless communication systems [1]. However, the THz region is still under exploration due to substantial physical limitations for both electron devices such as transistors, and photonic devices such as lasers [1, 2]. Low-power consumed, integrated, room-temperature operating novel solid-state THz laser devices is the critical demand for the brighter future society [3]. In such a situation, graphene, a monoatomic layer of sp^2-bonded carbons in a honeycomb crystal lattice, has drawn considerable attention due to its unique and superior carrier transport, optical, and plasmonic properties since it has been discovered [4–9].

Electrons and holes in graphene hold a linear dispersion relation with zero bandgap, resulting in extraordinary features like massless relativistic Dirac fermions with back-scattering-free ultrafast transport [4–6] as well as the negative-dynamic conductivity in the THz spectral range under optical or electrical pumping [9–11]. The author's group succeeded in single-mode emission at 5.2 THz and 1–8-THz broadband amplified spontaneous emission at 100 K from fabricated graphene-channel laser transistors [12]. However, its gain and output intensity are thoroughly limited by the absorption coefficient of monolayer graphene below 2.3% [9, 11]. To break through the substantial limit, graphene Dirac plasmons (GDPs) open a pathway toward the realization of intense THz laser transistors operating at room temperatures with dry-cell battery [7, 13, 14]. The physics behind the GDPs is the current-driven instabilities giving rise to self-oscillation and coherent amplification in the THz electromagnetic spectra [15–18]. In this chapter, the theory and experiments for graphene-based plasmonic terahertz laser transistors are reviewed.

2. Graphene THz laser transistors under current-injection pumping

2.1 Theory and modeling

When the intrinsic graphene is optically pumped with a photon energy $\hbar\Omega$, where $\hbar$ is the reduced Planck constant and Ω is the angular frequency of photon, interband transitions lead to the generation of "hot" photoelectron-photohole pairs in the distance equal to the photon energy at above and below the Dirac point for the electrons and holes, respectively (**Figure 1**). At room temperature, these photo-excited "hot" electrons and "hot" holes are quasi-equilibrated with the low-energy conduction electrons and valence holes around the Dirac point at the ultrafast time scale of tens of femtoseconds due to the carrier-carrier scattering resulting in the quasi-Fermi distribution (**Figure 1**) [11, 19]. As a result, the quasi-Fermi level of electrons and holes split and shift *below* and *above* the Dirac point, respectively. At the same time, the high-energy electrons and holes, lose their energy by emitting optical phonons. Thus the quasi-Fermi distribution, which is originally widely spread in the energy space, gets concentrated around the Dirac point. This results in a rapid recovery of the quasi-Fermi level on a picosecond time scale [19]. After the energy relaxation via the optical phonon emissions, the non-equilibrium electrons and holes may lose their energies via the "interband" optical phonon scattering, impurity–/disordered scattering, as well as the acoustic phonon scattering, or they recombine. Compared to the aforementioned "intraband" optical phonon scattering, these scattering and recombination processes have much longer relaxation times. Therefore, the nonequilibrium carriers are piled up around the Dirac point if the pumping intensity is sufficiently high. This is the carrier population inversion, in which the quasi-Fermi levels for electrons and holes shift *above* and *below* the Dirac point, respectively (**Figure 1**). The Auger-type three-particle scattering processes are theoretically forbidden in the ideal graphene [20], but they might dominate in the disordered low-quality graphene and/or under intense pumping leading to significant many-body effects [21]. Thus, the Auger-type scattering is recognized as a killer of the carrier

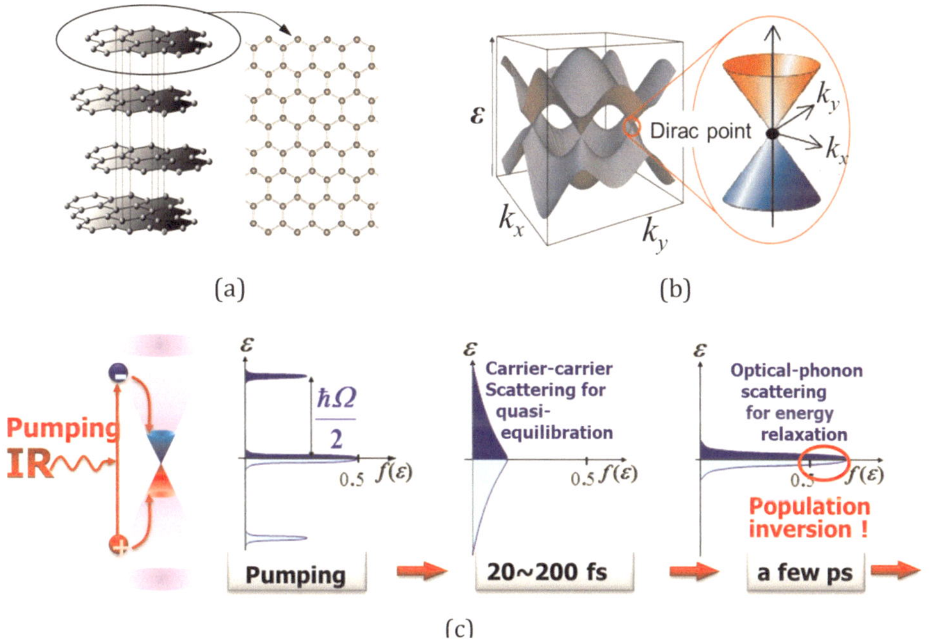

Figure 1.
Graphene and its carrier energy relaxation dynamics under optical pumpinp. (a) Real-space lattice image from bulk graphite (left) to graphene (right). (b) Energy band structure of graphene. (c) Carrier energy relaxation dynamics of intrinsic graphene under optical pumping. When graphene is optically pumped with a photon energy $\hbar\Omega$ photoelectron-photohole pairs are generated at the level $\hbar\Omega/2$ above and below the Dirac point, and then followed by ultrafast carrier-carrier scattering and then optical phonon scattering. Adapted with permission by IEEE from Ref. [11].

population inversion [22]. However, the carrier population inversion in the THz frequency range has been experimentally observed in prolongated long time more than 1 ps after a rather intense pumping with a high photon-energy (~1.5 eV) picosecond laser pulses in high-quality epitaxial graphene even at room temperature [23]. This result encouraged us to explore the creation of the graphene THz lasers/amplifiers.

Optical pumping suffers from significant heating of the electron-hole plasma suppressing carrier population inversion. The electron-hole plasma, however, can be cooled down if the pumping photon energies are sufficiently low. Recently rigorous theoretical modeling and calculation of the Auger scattering rate including many-body effects of Coulomb interactions reveal that the Auger scattering is substantially suppressed when carrier temperature is maintained at or below room temperature [24]. In this situation, current injection pumping is the idealistic way to suppress the carrier heating decreasing the pumping threshold because the pumping energy could be of the order of millielectron volts in a graphene p-i-n junction structure [10, 11].

A dual gate p-i-n structure using a graphene-channel field effect transistor (DG-GFET) has been proposed as the simplest current-injection THz graphene laser structure (**Figure 2**) [10, 12]. The conductivity profiles calculated for the typical dimensions and applied bias conditions demonstrate the advantage of current-injection pumping compared to optical pumping. The optical conductivity of the graphene channel under the gates and drain biases is given by [10, 12]:

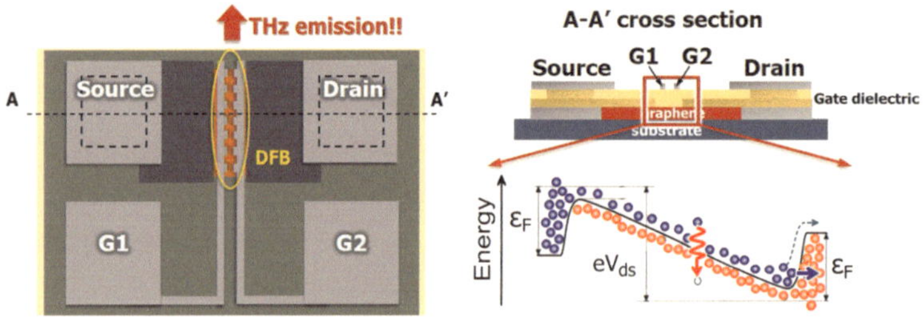

Figure 2.
DG-GFET (dual-gate graphene-channel field effect transistor). Top view (left), cross-sectional view (right upper), and a band diagram under complementary dual gate bias and forward drain-source bias condition getting population inversion and resultant spontaneous THz emission (right lower). A distributed feed-back (DFB) laser cavity is incorporated in the toothbrush-shaped dual gate electrodes in which the distance between the gates is periodically modulated to modulate the gain coefficient, producing resonant modes to the THz photons propagating along the DFB cavity (in plane DG finger direction). The DFB period gives the resonant mode frequencies and the modulation depth and the number of periods give the quality factor of the cavity [12].

$$\sigma_\omega = \sigma_\omega^{\text{intra}} + \sigma_\omega^{\text{inter}}, \tag{1}$$

$$\sigma_\omega^{\text{intra}} \approx \frac{e^2 \varepsilon_F}{2\pi\hbar^2} \cdot \frac{\tau}{1+\omega^2\tau^2}, \tag{2}$$

$$\sigma_\omega^{\text{inter}} \approx \frac{e^2}{2\hbar} \cdot exp\left(\frac{eV - 2e\sqrt{V_F V_g}}{2k_B T}\right) \cdot \sinh\left(\frac{\hbar\omega - eV_{ds}}{2k_B T}\right), \tag{3}$$

where $\sigma_\omega^{\text{intra}}$ and $\sigma_\omega^{\text{inter}}$ are the intraband and interband components of the conductivity, respectively, e is the elementary charge, ε_F is the Fermi energy reflecting the carrier doping level by applying V_g, τ is the carrier momentum relaxation time, k_B is the Boltzmann constant, and T is the temperature [10]. The gate biasing ($\pm V_g$) controls the injection level, whereas the drain bias (V_{ds}) controls the level of population inversion (the amount of shifting the quasi-Fermi levels of electrons and holes). To minimize the undesired tunneling current lowering the injection efficiency the distance between the dual gate electrodes should be sufficiently long. When the carrier momentum relaxation time is rather long ($\tau = 2.0$ ps) and carrier injection level is pertinent at V_g = 1.0 V, an even weak V_{ds} produces a net gain (negative conductivity) in a wide THz frequency range (**Figure 3a**) [12]. The situation dramatically changes when the value of τ becomes short (down to 0.1 ps); in this case, the net gain almost disappears (**Figure 3b**) [12]. As a consequence, the high-quality graphene having a picosecond-order value of τ is mandatory to obtain a net gain in the THz range. It is noted that the maximal available negative conductivity is limited below $e^2/(4\hbar)$ corresponding to the interband absorption coefficient ($\alpha = \pi e^2/(c\hbar) \approx 2.3\%$) of monolayer graphene [9].

2.2 Experiments

The author's group designed and fabricated the DG-GFETs incorporating a distributed feed-back (DFB) type internal cavity, demonstrating the world-first observation of single-mode lasing from GFETs at 5.2 THz at 100 K (**Figure 4**) [12]. The graphene used in the GFETs is non-Bernal-stacked a few layers of high-quality graphene synthesized by the thermal decomposition of a C-face 4H-SiC substrate. The GFET was fabricated using a standard photolithography and a gate stack with a SiN

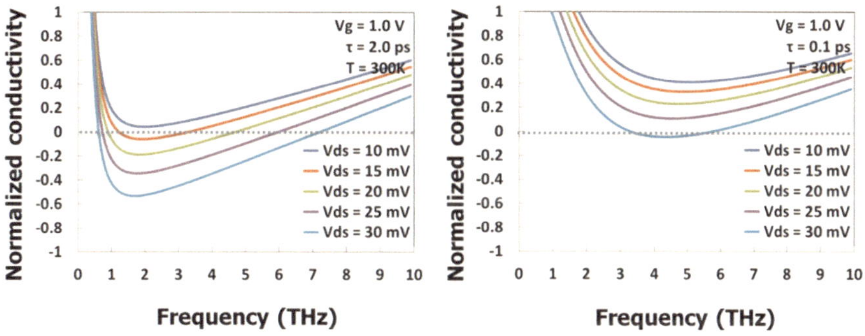

(a) (b)

Figure 3.
Real part of the conductivity spectra of a complementary dual-gate bias DG-GFET at 300 K with varying drain-source bias voltage. The vertical axis is normalized by the fundamental conductivity (e2/4h), which corresponds to 2.3% absorbance of monolayer graphene. Vg (= − Vg1 = +Vg2) = 1.0 V. (a) τ = 2.0 ps. (b) τ = 0.1 ps. Adapted with permission by the authors under the creative commons attribution CC-BY 4.0 international license from Ref. [12].

dielectric layer providing an excellent intrinsic field-effect mobility exceeding 100,000 cm^2/Vs at 300 K. A pair of toothbrush-shaped gate electrodes was patterned to form a DFB cavity in which the active gain area and corresponding gain coefficient were spatially and periodically modulated. The THz single mode lasing was observed at around 5.2 THz with an emission intensity as high as 0.1 μW and with a linewidth of 31 GHz under pertinent complementary DG biases ($\pm V_g$) and V_{ds} conditions serving the carrier population inversion in good agreement with numerical simulations (**Figure 4a**). A different sample with the identical design on the same wafer exhibited an amplified spontaneous broadband emission ranging from 1 to 7.6 THz at 100 K with a maximal emission power of 80 μW (**Figure 4b**).

2.3 Discussion toward introduction of graphene plasmons

Figure 4 shows the variation of THz emission characteristics stems from a poor THz photon field confinement and resultant weak gain overlapping. Numerical analysis well reproduces such a variation when a wide fraction of the carrier momentum relaxation time of graphene carrier under weak gain overlapping conditions is assumed [12, 25]. Improvement in the gain overlapping by introducing a plasmonic waveguide for dense THz photon field confinement as well as on the cavity quality factor by increasing the number of the DFB periods and the DFB modulation depth will lead to more intense lasing at higher temperatures approaching 300 K [25]. Also, the introduction of the GDP dynamics is a promising way to dramatically increase the THz net gain and quantum efficiency of the GFET lasers. Due to the extremely slow-wave nature (by two orders slower than the speed of light) of the GDPs [7, 8]. Once THz photons couple with the graphene plasmons, photons may propagate along the population-inverted graphene as plasmon polaritons having the plasmon velocity so that the interaction between the THz photons and graphene carriers is dramatically enhanced in proportion to the ratio of the speed of photon to that of the plasmon, well exceeding the aforementioned quantum-mechanical limit of 2.3% per layer of graphene [26]. A giant gain enhancement of up

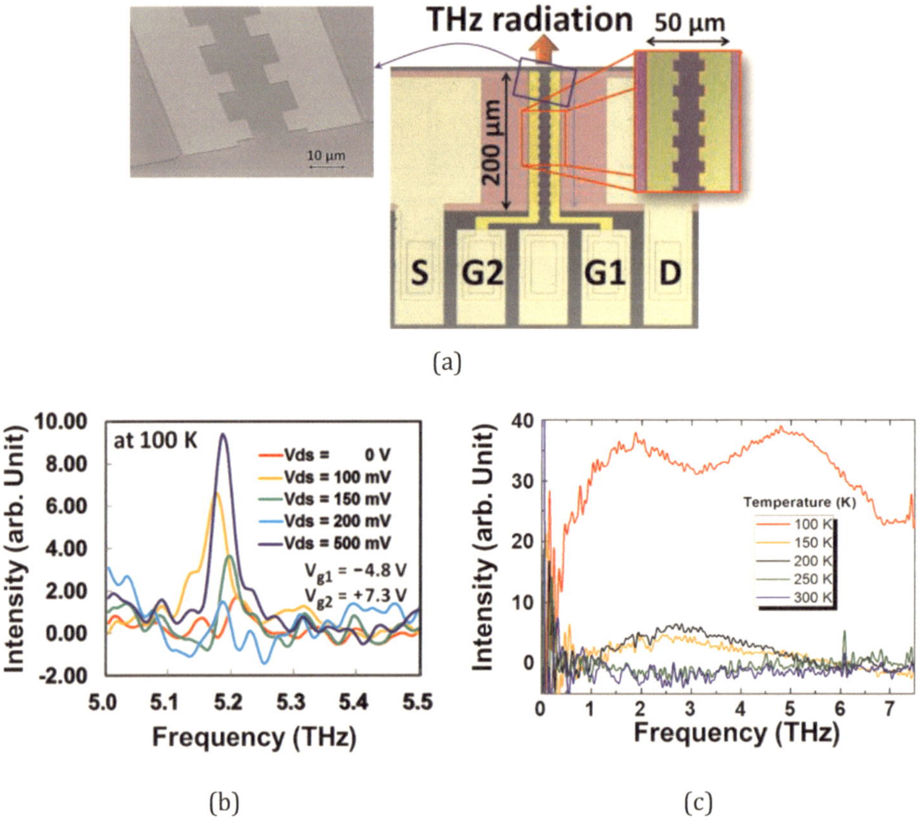

Figure 4.
Measured emission spectra from fabricated DFB-DG-GFET samples. (a) Electron scanning and optical microphotographic images of a fabricated sample. (b) Single-mode emission at ~5.2 THz at 100 K with a ~ 0.1 μW and a minimal linewidth of ~31 GHz from one fabricated DFB-DG-GFET sample. (c) Amplified broadband spontaneous emission in a frequency range from 1 to 7.6 THz at 100 K from a different DFB-DG-GFET sample. The threshold operating temperature stays in between 100 and 150 K. Adapted with permission by the authors under the Creative Commons Attribution CC-BY 4.0 International license from Ref. [12].

to 50 times as high as that without graphene plasmons has been experimentally observed in the THz range [27].

3. Graphene-plasmonic instabilities for THz amplification

3.1 Theory and modeling

Unlike plasmons in the bulk, two-dimensional plasmons have a wavenumber dependence, and when they are localized only by an in-plane crystal field on a two-dimensional surface (non-gate-controlled plasmons) or electrostatically by an external vertical electric field (gate-controlled plasmons), as in a transistor gate structure, their dispersion relations are different [7, 28]. Furthermore, the plasmon dispersion relation is different between ordinary two-dimensional electron systems, which have a quadratic dispersion relation due to the existence of electron effective mass, and Dirac two-dimensional electron systems, which have linear dispersion due to the disappearance of electron effective mass, such as electrons in graphene [28]. However, for the electron concentration, the plasmon energy is proportional to the 1/4 power of the electron concentration due to the peculiarity of the linear dispersion of the Dirac electrons [7, 28]. When plasmon is localized in the element structure, the plasmon mode stands at a specific frequency determined by the structural dimensions and the plasmon velocity, and if the structural dimensions are less than or equal to the average free pass of the electrons, the plasmon coherency is conserved and plasmon resonance is promoted.

When plasmon is excited in a two-dimensional electron system traveling in direct current within a limited region such as the channel in a transistor, the difference between the electron drift velocity and the plasmon velocity modulates the spatio-temporal density of the plasmon. At the boundary edges of this region, although the electron travel is guaranteed its continuity by the current continuity condition, the plasmon (plasma wave) is subject to reflection due to impedance mismatch at the boundaries. As a result, a standing wave is generated by the superposition of the forward-traveling waves and the reflected backward waves. When the singular or multiple wavelength(s) of the plasma waves meet the channel length, the amplitude of the plasmon increases divergently with each repeated reflection, and the plasmon falls into an unstable state. This phenomenon is called plasmon instability and is a self-excited oscillation phenomenon that occurs when the DC energy of the electron is converted to plasmon energy near the resonance frequencies [15, 28].

Plasmon instability has multiple mechanisms for its occurrence, depending on the relationship between the electron drift velocity v_d and the plasmon velocity s. Under the condition $v_d < s$, which is generally established, there are two types of instability; Dyakonov-Shur (D-S) instability [15] and Ryzhii-Satou-Shur (R-S-S) instability [16]. In case of the D-S instability, the plasma wave amplitude increases with each reflection at the open end where the reflection coefficient is close to 1 due to the Doppler shift effect associated with the asymmetry of the plasmon boundary, whereas the R-S-S instability is caused by electron bunching effects due to electrons traveling at different drift velocities inside the plasmon resonator at high electron concentration and outside the resonator at low electron concentration. At the boundary where the drift velocity exceeds the plasmon velocity, $v_d > s$, a plasmon shock wave type instability (plasmonic boom instability) [17] as well as the R-S-S type, and furthermore, within a periodic electron concentration modulation structure, a Cherenkov-type instability, where the plasmon bunching is divergently enhanced, are observed, respectively [17].

We have proposed a graphene-channel field-effect transistor with an "asymmetric dual-grating-gate" (ADGG-GFET), in which a broadband antenna mechanism that efficiently couples terahertz electromagnetic waves and graphene plasmon are woven into the transistor electrodes [29]. The ADGG consists of two independent gate electrodes, G1 and G2, arranged in an interdigitated manner (**Figure 5a**). A high bias is applied to one G1 to accumulate highly concentrated electrons in the channel region below G1, and a Dirac potential is applied to the other G2 to deplete the channel region below G2, periodically modulating the electron concentration in the channel (**Figure 5b**). In this situation, when a DC bias is applied to the drain, the electron drift velocity is also periodically modulated in response to the spatial distribution of the electron concentration in the channel to preserve the current-continuity condition. As a result, the plasmon instability described above is excited. The high electron concentration region functions as a plasmon resonator, leading to the self-excited oscillation of plasmon near the plasmon resonance frequency. The plasmon itself is a

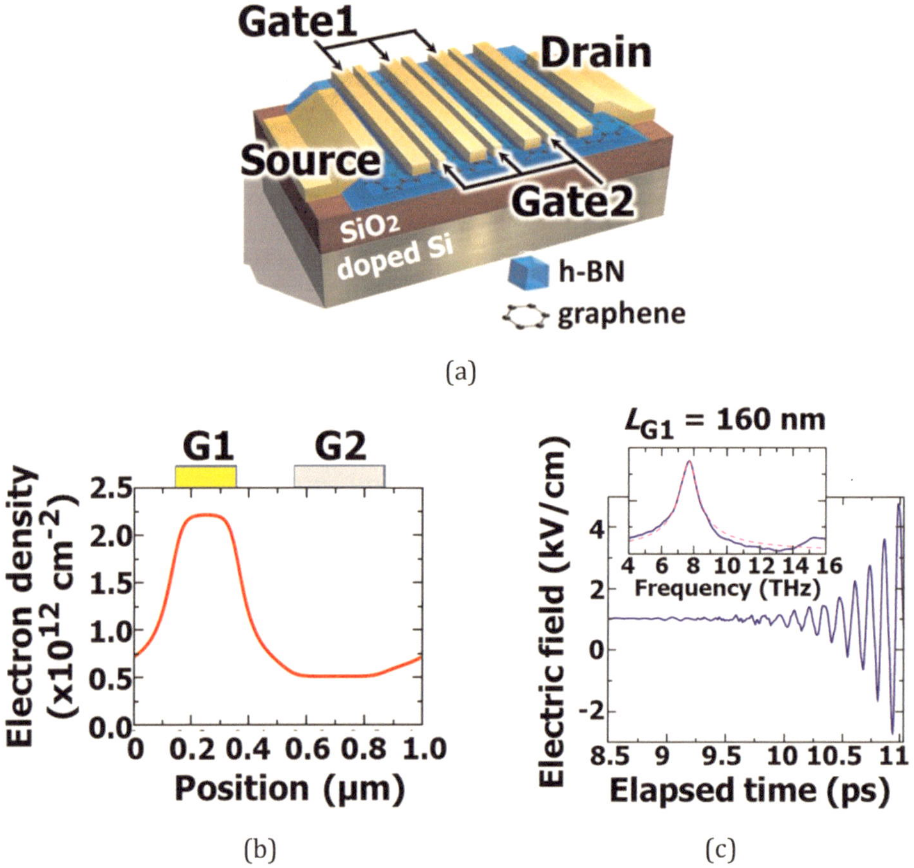

Figure 5.
An example of numerical analysis of plasmon instability in an ADGG-GFET. (a) A bird's eye schematic of an outlook of the ADGG-GFET. (b) Spatial density modulation of electrons in a unit section of the ADGG structure. The high-electron-density region underneath G1 works as a plasmonic cavity whereas the low-electron-density region underneath G2 serves a high-speed electron transit region and works to excite plasmons in the cavity. Adapted with permission by American Physical Society from Ref. [30]. (c) Numerically simulated temporal evolution of the longitudinal electric field intensity after turning on the DC drain bias beyond the threshold level, leading to coherent self-oscillation at the frequency of plasmon resonance determined by the cavity length L_{G1} *and the electron densities in the cavity. Adapted with permission by American Physical Society from Ref. [30].*

non-radiative evanescent wave, but it is converted into an electromagnetic wave through the antenna mechanism of ADGG, leading to free-space radiation. If the plasmon cavity length (width of each finger) of the gate grating electrodes is set to submicron dimensions, self-excited oscillation in the THz band can be obtained. If a THz wave with a frequency component near the oscillation frequency is injected from the outside, the injection locking synchronization phenomenon will be activated and the self-excited oscillation will coherently synchronize with the injected terahertz wave. As a result, an amplification of the stimulated emission of the incident terahertz wave is obtained. Therefore, it can be said that the graphene plasmon instability can cause coherent resonant amplification of electromagnetic waves. We have successfully modeled the plasmonic electromagnetic response of the ADGG-GFETs and numerically verified the occurrence of THz self-excited oscillation phenomena originating from graphene plasmon instability (**Figure 5c**) [30].

3.2 Experiments

To demonstrate the light amplification of stimulated emission of THz radiation by excitation of graphene plasmons as described above, we fabricated a prototype ADGG-GFET with the help of 2-DTech Corp. Graphene sheets of the highest quality and hexagonal boron nitride (hBN), which is widely used as a chemically inert and mechanically stable material that less deteriorates graphene carrier transport, were deposited on bulk graphite and bulk hBN materials, respectively, by a peeling and transfer method and then sequentially. The hBN layer completely seals the top and bottom surfaces of the graphene channel. Source and drain electrodes were formed by edge-junction with graphene channels, whereas the gate electrodes were formed by electron beam lithography and a lift-off method. The amount of doping of the entire graphene channel can be back-gate controlled from the backside of the substrate. The finger length of the gate, which determines the plasmon resonance frequency, was differentiated between the two gates of the ADGG, and by selecting the gate that accumulates electrons when a high bias is applied, two resonator lengths can be evaluated in a single FET. Two prototype FETs with two different characteristic dimensions were evaluated for a total of four resonator structures C1 to C4 with different dimensions of 0.5, 0.75, 1.0, and 1.5 μm (**Figure 6a**). Measurements of the DC transfer characteristics confirmed the good ambipolar properties characteristic of bandgap-less graphene. The field-effect mobility extracted from the DC properties showed excellent values above 30,000 cm^2/Vs [13, 14].

The temporal response of the transmitted wave of the prototype ADGG-GFET to an incident terahertz pulse was measured by varying the DC voltage applied to the drain terminal using terahertz time-domain spectroscopy (TDS). To excite graphene plasmon resonance, in which the resonator structure is defined along with the shorter dimension corresponding to the gate finger of ADGG, it is necessary to align the polarization axis of the incident THz pulse in the channel direction. We controlled the incident terahertz pulse to be linearly polarized and verified the effect of plasmon excitation for two conditions: parallel and orthogonal polarization axes in the channel direction. All experiments were performed at room temperature.

First, the THz pulse response of ADGG-GFETs was measured with no drain bias applied. As a result, a strong resonant absorption property with an absorption rate exceeding a maximum of 16%, originating from a clear plasmon resonance, was measured only when the polarization axis of the incident terahertz pulse was aligned in the channel direction. When the polarization axis was orthogonal to the channel direction, a Drude absorption spectrum associated with free carrier absorption with a monotonically decreasing absorption rate with increasing frequency was observed (**Figure 6b**).

Next, the same experiment was conducted by gradually increasing the drain bias. As the drain bias increased, the resonance absorption decreased with a redshift (decrease in peak frequency). When the drain bias reached the first threshold level, the absorption turned to be fully transparent over the entire observed frequency band. When the drain bias was further increased to reach the second threshold, it turned into a resonant amplification characteristic. As the drain bias was further increased, the resonant amplification showed a blueshift (increase in peak frequency), and the amplification ratio increased up to 9% (**Figure 6c**). Asynchronous components such as self-excited oscillations associated with plasmon instability, for example, are not detected in principle. Therefore, the obtained amplification effect can be understood as a coherent light amplification of stimulated emission of the incident THz wave

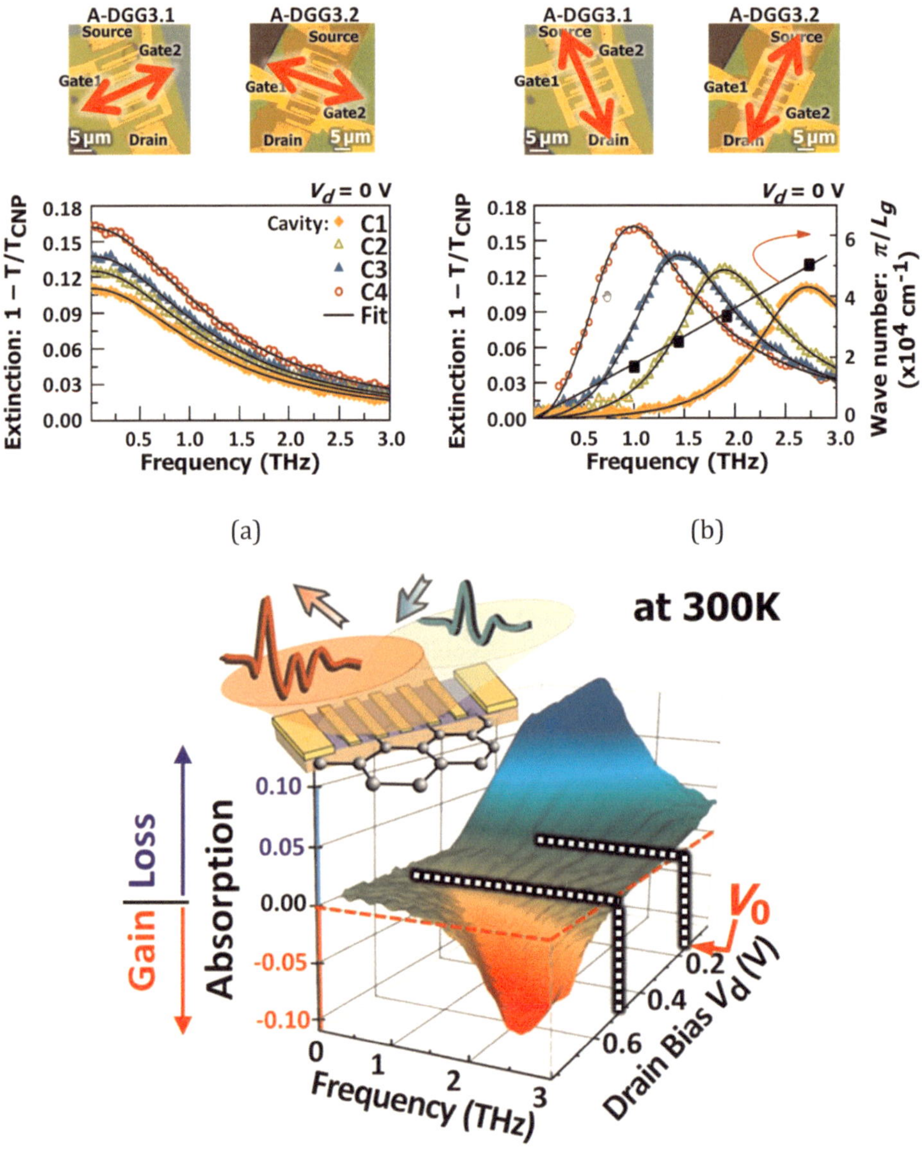

(a) (b)

(c)

Figure 6.
Experimental results of THz temporal and frequency response of fabricated ADGG-GFETs using a TDS method. (a) Electron microphotographs of fabricated samples and their Drude-like absorption spectral characteristics under drain-unbiased conditions where plasmon is not excited with incident polarization (designated with red arrows) perpendicular to the ADGG finger. Adapted with permission by the authors under the Creative Commons Attribution CC-BY 4.0 International license from Ref. [13]. (b) Electron microphotographs of fabricated samples and their resonant absorption spectral characteristics under drain-unbiased condition where plasmon is excited with incident polarization (designated with red arrows) parallel to the ADGG finger. Adapted with permission by the authors under the Creative Commons Attribution CC-BY 4.0 IInternational license from Ref. [13]. (c) Resonant absorption and amplification spectral responses when drain biased is varied from 0 up to ~1 V.

radiation, in which the DC power energy supplied by the application of a DC drain bias is converted into an induced amplification component via the onset of plasmon instability.

The redshift observed in resonant absorption can be understood as the normal Doppler effect since the electron drift velocity increases with drain bias, while the blueshift observed in resonant amplification can be understood as the reverse Doppler effect. This effect means that the frequency response can be actively modulated and controlled by graphene, which is a discovery of a potential mean of new terahertz-band signal processing functionalities with a significant impact on the related engineering and industrial fields.

The experimental results of up to 16% absorption and 9% amplification of graphene greatly exceed the upper limit of quantum efficiency obtained by the direct interaction of photons of incident electromagnetic waves with graphene electrons, i.e., 2.3%, which corresponds to the optical absorption associated with direct interband transitions. The drain bias dependence of the obtained resonant dispersion properties is in good agreement with the properties in the plasmonic boom and Cherenkov-type instability that have been theoretically clarified. On the other hand, the drift velocity of graphene electrons with drain bias and the plasmon velocity were extracted, and the drift velocity was always lower than the plasmon velocity, suggesting that the operating mechanism of the plasmon instability is the expression of the D-S type and R-S-S type instabilities and that the plasmonic boom and Cherenkov-type instability were ruled out. In response to this contradiction, we newly modeled the plasmon dynamics of ADGG-GFETs and analyzed their properties, and found that the resonance dispersion properties observed even under conditions where the drift velocity does not exceed the plasmon velocity could be caused by the R-S-S type instability [13]. It is also suggested another possibility of the occurrence of a different type of new instability mechanism called Coulomb-drag instability of GDFs [31–33] of GDFs which we have recently discovered.

3.3 A discovery of a new instability mechanism in GFETs

Recently, we have discovered a new instability operating mechanism based on the Coulomb drag interaction of graphene Dirac plasmons [31–34]. The upper limit of the kinetic velocity of graphene Dirac fermions (GDFs) is defined by the slope of the band linear dispersion, i.e., the Fermi velocity $v_F\left(\approx 10^8\ \mathrm{cm/s}\right.$ even at RT$\left.\right)$, which enables extremely high and micron-order long-distance ballistic transport even at room temperature [35]. At the same time, GDFs have extremely high viscosity, higher than that of honey, and the electron-to-electron scattering due to Coulomb interaction is extremely strong [36]. Due to these unique properties of GDFs, which are separated from the massive electrons in existing semiconductors, in the GFET structure with a channel length of less than a micron order under consideration, if GDFs are injected from the source end, they are transported in a ballistic manner in the ungated intrinsic region and reach the gate region with extremely high kinetic energy with the velocity of v_F. As a result, the quasi-equilibrium electrons (QEs) that aggregate in the gate region undergo strong Coulomb interaction with the ballistic transport electrons (BEs). Together with the high viscosity property of the GDFs, they are transformed into quasi-equilibrium drag electrons (DQEs) that enter a so-called drug state (**Figure 7a** and **b**) [31]. If the drain bias is applied weakly to the extent that the potential slope between the gate and drain is inverted from the slope of the forward potential between the drain and source, the inverted potential slope will cause QEs to be injected back into the channel from the drain end as depicted in **Figure 7a**. This has

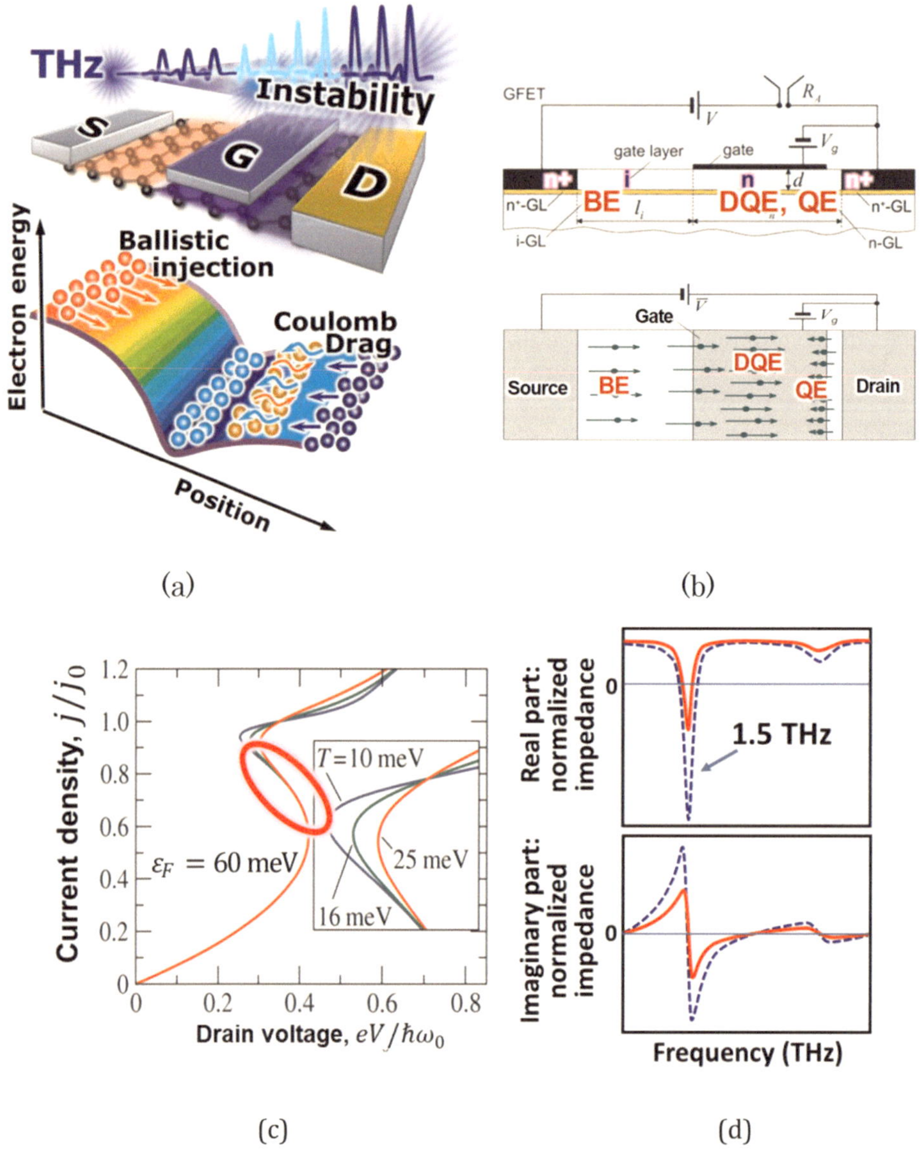

Figure 7.
Coulomb-drag instability of graphene Dirac plasmons. (a) Bird's eye view of the device structure and related band diagram under a sufficient gate bias and a pertinent drain bias condition to promote the instability. Adapted with permission by Wiley-VCH GmbH from Ref. [32]. (b) Cross-sectional device structure and carrier flow. BE: ballistic electron, QE: quasi-equilibrated electron, DQE: dragged quasi-equilibrated Electron. (c) Typical channel current versus drain bias voltage at a gate-bias-dependent Fermi level of 60 meV for different temperatures of 10, 16, and 25 meV. Adapted with permission by American Physical Society from Ref. [31]. (d) Typical impedance when the Coulomb-drag instability is promoted in a GFET. The vertical axis is normalized to the i-region resistance. The fundamental plasmon resonant mode is characterized by the gated channel length and the electron density in the gated area at 1.07 THz. Adapted with permission by Wiley-VCH GmbH from Ref. [32].

the effect of reducing the channel current as the drain potential rises, resulting in negative differential conductivity (**Figure 7c**) [31, 32].

The channel currents consist of ballistic transport electron flow BE, quasi-equilibrium electron flow QE, and quasi-equilibrium drag electron flow DQE by

GDFs. Under such a situation, the current continuity was modeled by Kirchhoff's law, and the DC current-voltage characteristics and AC impedance were analytically obtained [31–33]. As a result, we theoretically discovered that, first, assuming micron-order GFET structure dimensions, extremely strong negative differential conductivity characteristics can be obtained in a wide THz band even at room temperature (**Figure 7c**) [31], and that the GFET gate region functions as a graphene plasmonic coherent oscillator in the THz frequency range (**Figure 7d**) [32, 33].

Second, in dual-gate GFETs, if a p-i-n junction is formed in the channel by applying complementary gate bias, the slope of the p-i-n junction with the application of drain bias becomes linear and steep, resulting in the valence electron holes in the p region and conduction electrons in the n region forming a Zener Klein tunneling [37], resulting in strong Coulomb drag as ballistic transport electron holes, and that this graphene tunneling transistor (GTT) also exhibits strong GDP Coulomb drag instability comparable to or stronger than that of single-gate GFETs (**Figure 8**) [34]. Moreover, the discovered instability has no high-speed requirement for carrier velocity in the plasmon resonator region, and the instability index under real device conditions is several times higher than other instabilities currently known (the Doppler-shifted D-S type, the Cherenkov-boom type, and the electron velocity modulation R-S-S type), and the prospect of achieving THz radiation intensity of 100 μW or higher under room temperature has been analytically demonstrated [34].

The Coulomb drag effect has been originally considered in a double-layered graphene structure in which two independent graphene are separated by an atomically thin insulation layer and electrostatically coupled working as a graphene nanocapacitor [38]. Our finding that the drag effect can take place even in a single graphene layer brings the effect into a simple GFET structure, diversifying its functionalities. In principle, the Coulomb drag mechanism can promote the plasma wave generation in the periodic GFET structures analogous to the previously considered two-stream instability systems with ballistic carrier flows and quasi-equilibrated non-traversing carriers in massive compound semiconductor heterostructures [39]. However, these classical massive 2D electron systems cannot totally mediate such a strong Coulomb drag effect as in GDFs due to their poor carrier transport properties. In summary, in the proposed GFET system, the critical differences and advantages over previous implementations are (i) the extraordinary transport properties of GDFs, which allow for ballistic transport over micrometers in graphene channels even at room temperature, and (ii) the strong Coulomb drag effect via the strong Coulomb inter-carrier scattering of GDFs.

3.4 Actively controlling the PT-symmetry of graphene Dirac plasmons

In quantum mechanics, it was discovered that even non-Hermitian Hamiltonians can provide realistic solutions if the system preserves the parity and time-reversal (PT) symmetries [40]. The author introduces active controlling of the PT symmetry to our original DGG-GFET structure by applied voltages and demonstrates a 100-Gbit/s-class ultrafast gain switching operation [11].

The PT symmetry can be expressed by a pair of complementary gain and loss mediums (**Figure 9**), leading to a perfect transmission property (unidirectionality) [41]. We first numerically confirmed such an anisotropic wave propagation in a PT-symmetric plasmonic metasurface by using a simple toy model of a transmission line consisting of a series of the unit kinetic-inductance/electrostatic-capacitance/Drude-conductance circuitry (**Figure 10a**), clearly reproducing the unidirectionality

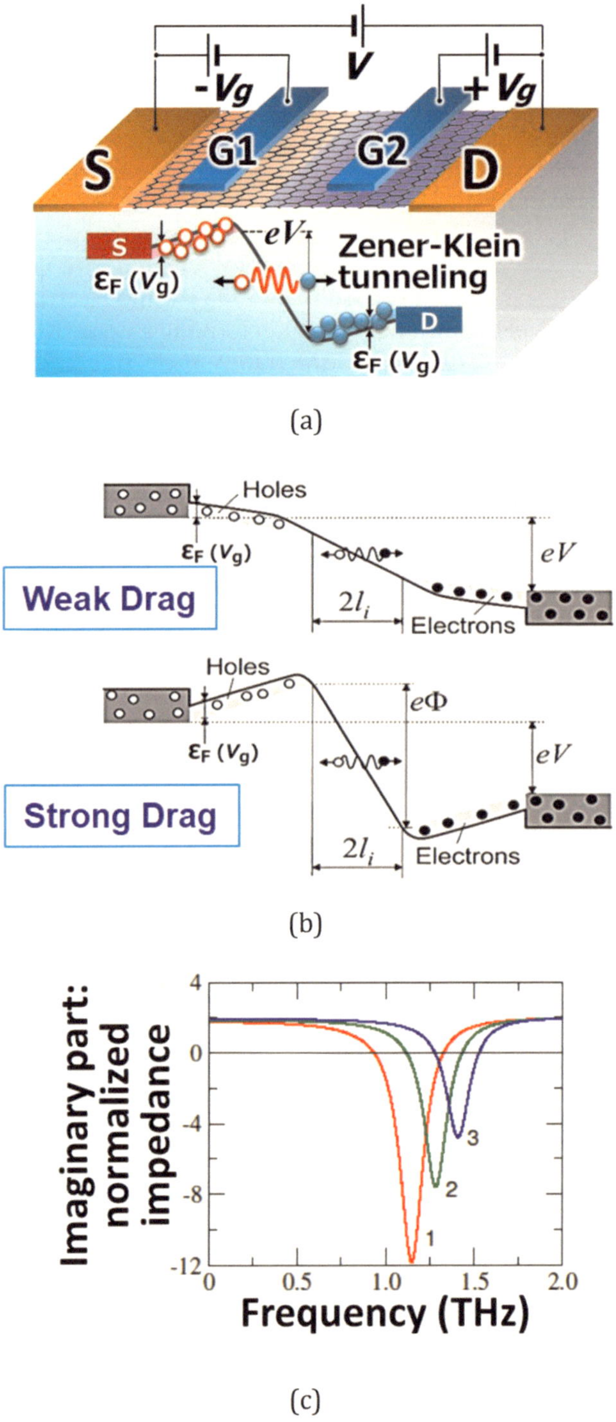

Figure 8.

Promotion of Coulomb-drag instability in a tunnel GFET. (a) Bird's eye view of the tunnel GFET structure and its band diagram under a strong drugged condition. (b) The change of the band diagram from a weak to a strong drag state. (c) Numerically calculated real part of the impedance normalized to the i-region resistance. Plots no. 1, 2, and 3 are the cases for the gate-bias-dependent plasmon mode frequency of 1.15, 1.3, and 1.4 THz, respectively. Adapted with permission by American Physical Society from Ref. [34].

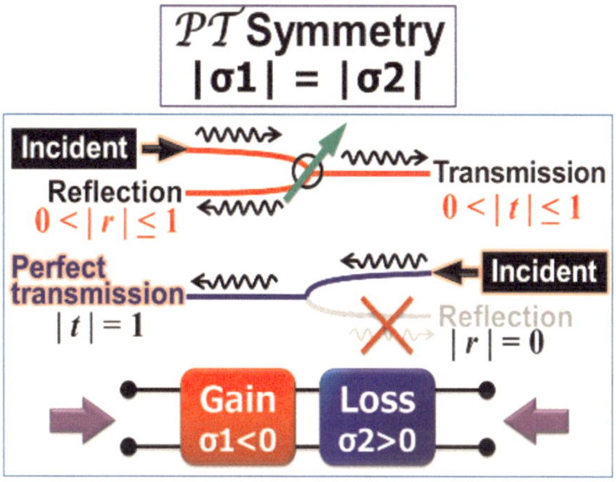

Figure 9.
Expression of the PT symmetric system in an electrically equivalent two-port circuit with a series of complementary gain and loss elements, demonstrating its unidirectionality of electromagnetic wave propagation having a normal transmission/reflection transmission when the signal is incident from the gain port and a perfect transmission when the signal is incident from the loss port. Adapted with permission by the authors under the Creative Commons Attribution CC-BY 4.0 International license from Ref. [11].

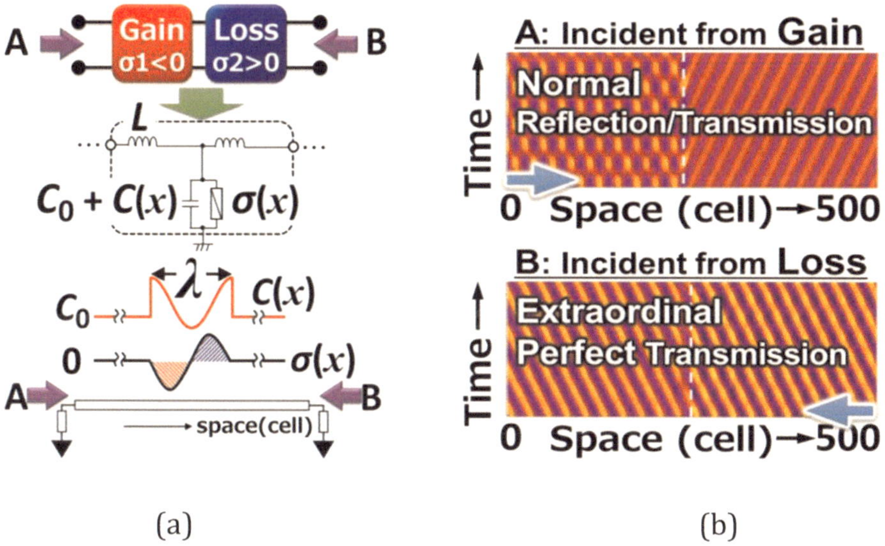

Figure 10.
Transmission line modeling of a PT-symmetric system and numerical analysis for reproducing is a unidirectionality characteristic. Adapted with permission by the authors under the Creative Commons Attribution CC-BY 4.0 International license from Ref. [11]. (A) A toy model for a PT-symmetric system in a distributed transmission line circuitry whose central core unit is given by the non-zero sinusoidal complementary gain-loss conductance element. (B) Spatio-temporal phase mapping of propagating the electromagnetic wave, demonstrating anisotropic, extraordinary perfect transmission when radiation is incident from the loss port.

(**Figure 10b**). When the GDPs in a DGG-GFET become unstable under forward-drain-biased, complementary DGG-biased conditions, the GDP metasurface is represented by a series connection of unit cells of the instability-driven gain section and the carrier-depleted loss section (**Figure 11**).

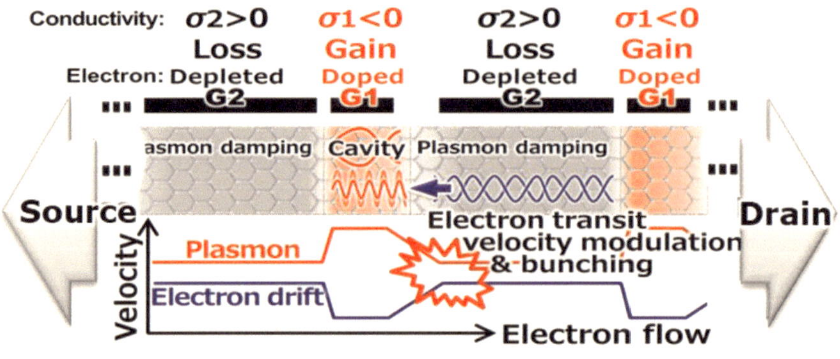

Figure 11.
Implementation of active control of the PT symmetry in a DGG-GFET in a case of R-S-S type GDP instability to be promoted. Adapted with permission by the authors under the Creative Commons Attribution CC-BY 4.0 International license from Ref. [11].

We numerically analyzed the transient response of a DGG-GFET to reveal how fast its laser cavity Q values are electrically controllable by using self-consistent simulation based on quasi-classical Boltzmann equation (**Figure 12**) [30]. The drain bias V_d was turned on at time zero. The electric field intensity, which is mainly applied to the depleted electron-transit region, was set at 0.8 kV/cm. The equivalent electron velocity

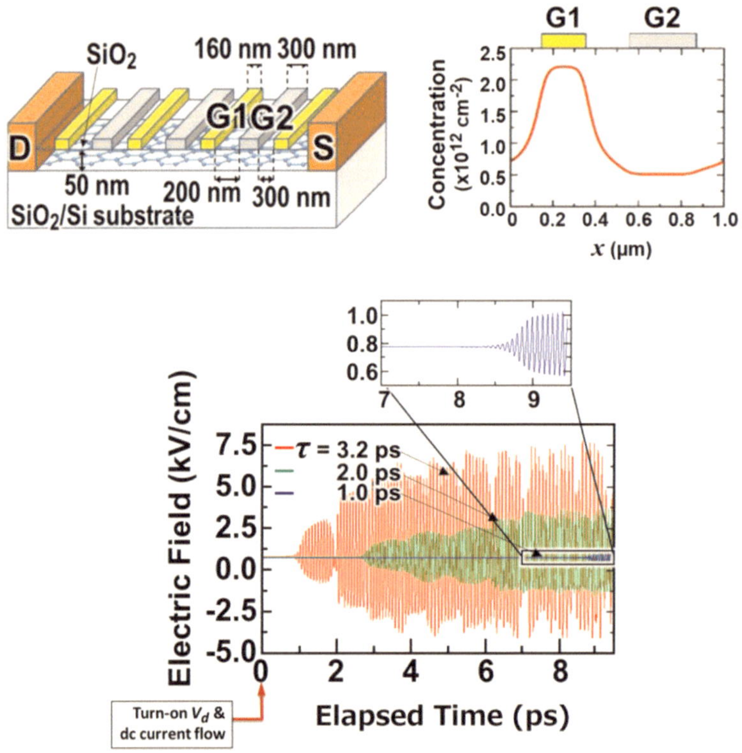

Figure 12.
Numerically simulated temporal evolution of GDP field activated by the dc-current-driven PT symmetry turned on at t = 0 ps for different momentum relaxation times τ of GDFs reaching the self-oscillation at the GDP mode frequency in a picosecond time scale. Adapted with permission by the authors under the Creative Commons Attribution CC-BY 4.0 International license from Ref. [11].

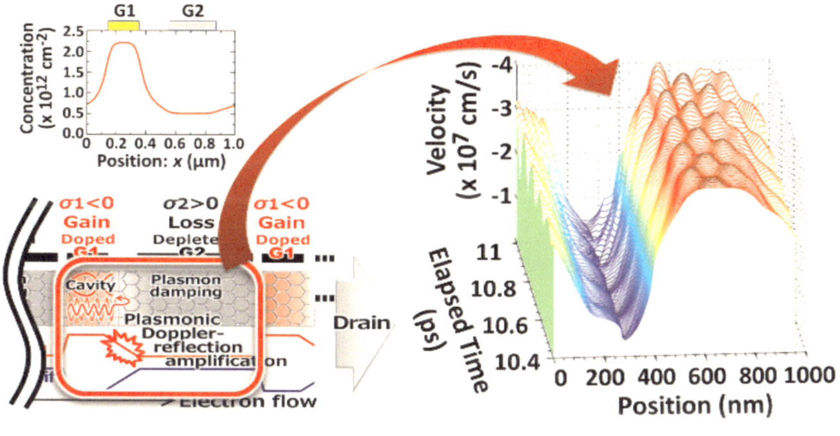

Figure 13.
Numerically simulated, temporally evolved spatial distribution of the carrier velocity in the DGG metasurface under THz self-oscillation promoted by the current-driven GDP instability. Adapted with permission by the authors under the Creative Commons Attribution CC-BY 4.0 International license from Ref. [11].

is estimated to be 4×10^7 cm/s at the field-effect mobility of 50,000 cm^2/Vs. **Figure 12** plots the temporal evolutions of the longitudinal electric field in the GDP cavity for τ = 1.0, 2.0, and 3.2 ps. Originally the graphene metasurface with zero-V_d is entirely lossy. With increasing V_d and the level of current injection pumping the GDP instability started to be promoted as is demonstrated in **Figure 6c**. When the instability-driven gain becomes balanced to the loss in the depleted region, the system becomes PT-symmetric. Further increase of V_d makes the level of instability-driven gain surpass that of the loss, resulting in THz self-oscillation of radiation emission. The important feature is its ultrafast transition speed within 10 ps duration. As is seen in **Figure 12**, the superior property with a longer τ value exhibits faster turning on with stronger field intensity of THz self-oscillation. Even for a rather moderate τ value of 1 ps the result demonstrates still fast transition to the stationary stage oscillation within 10 ps corresponding to the single-bit time slot for the 100 Gbit/s data rate.

Figure 13 plots the temporal variations of the microscopic spatial distribution of the electron velocities in the unit cell of the GDP metasurface after more than 10 ps passed from the triggering where the field becomes the stationary state. The red-colored plotted region corresponding to the depleted region underneath G2 where GDFs travel fast clearly reproduces coherent oscillatory modulation, manifesting ignization of GDP-instability driven THz self-oscillation.

4. Possible scenario toward graphene plasmonic THz laser transistors operating at room temperature

Based on the previous discussion, the author proposes a possible design of a graphene plasmon THz laser transistor based on the DGG-GFET structure as depicted in **Figure 14**. Suppose DGG G1 and G2 are complementary biased for bipolar (electron and hole) injection and forward biased between drain and source. The proposed device consists of a series of unit sections including the photonic seed section and the plasmon gain section. The photonic seed section generates spontaneous THz photons by current injection pumping, and the plasmon gain section amplifies the spontaneous

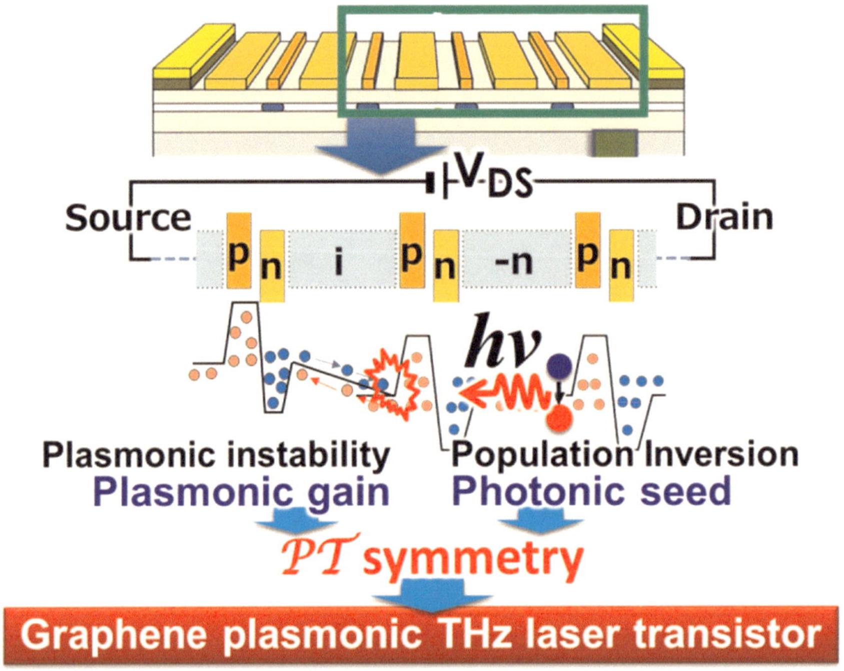

Figure 14.
A graphene plasmonic THz laser transistor structure of the author's proposal consisting of a series of unit sections including the photonic seed and the plasmonic gain sections. Adapted with permission by the authors under the Creative Commons Attribution CC-BY 4.0 International license from Ref. [11].

THz photons mediated by GDP instability. As described in Section 3.4, the idea of controlling the PT symmetry of the DGG-GFET metasurface helps to improve their performance and functionality such as ultrafast gain switching modulation.

The photonic seed section is selectively weakly doped (indicated by "-n" in **Figure 14**) in the wide ungated spacing of the photonic seed section to maintain a nearly flat band and retain inverted carriers in it under a slightly stronger drain bias voltage that promotes GDP instability in the plasmon gain section. This can be configured by selective weak n-type doping, but it may also be technically challenging. Electrostatic doping with back-gate biasing is a mature solution, but it is not easy with epitaxial graphene thermally decomposed on SiC substrates because it requires ultra-thin semi-insulating SiC substrates. Another method is chemical doping, but it has problems such as unintentional doping to other parts of the substrate, degrading the crystallinity, and shortening of carrier momentum relaxation time. Recently, a new technique has been developed to synthesize high-quality few-layer epitaxial graphene on single-crystalline SiC thin films grown on Si wafers [42]. Precise control of the epitaxially grown graphene layers also remains a challenge, which could be addressed by introducing microfabricated SiC substrate technology that can spatially limit the epitaxial area [43]. New ideas for such advanced devices and process technologies are the subject of future research.

5. Conclusions

In this chapter, the theory and experiments toward graphene-based plasmonic THz laser transistors were reviewed. The THz region is still under exploration due to substantial physical limitations for both electron devices such as transistors and photonic devices such as lasers. Low-power consumed, integrated, room-temperature operating novel solid-state THz laser devices are the critical demand for a brighter future society. This is the reason why graphene, a monoatomic layer of sp2-bonded carbons in a honeycomb crystal lattice, has drawn attention due to its unique and superior carrier transport, optical, and plasmonic properties since it has been discovered in the mid-2000s. Electrons and holes in graphene hold a linear dispersive energy spectrum with zero bandgap, resulting in extraordinary features like massless relativistic Dirac fermions with back-scattering-free ultrafast transport as well as the negative-dynamic conductivity in the THz spectral range under optical or electrical pumping. The author's group succeeded in single mode emission at 5.2 THz and 1–8-THz broadband amplified spontaneous emission at 100 K from current-driven graphene-channel laser transistors. However, its gain and output intensity are thoroughly limited by the absorption coefficient of monolayer graphene below 2.3%. To break through the substantial limit, graphene Dirac plasmons (GDPs)opened a pathway toward the realization of intense THz laser transistors operating at room temperatures. The physics behind the GDPs is the current-driven instabilities giving rise to self-oscillation and coherent amplification in the THz electromagnetic spectra. The author's group designed and fabricated ADGG-GFETs and demonstrated room-temperature THz amplification with a maximum gain of 9% which was four times larger than the quantum mechanical limit when THz photons directly interact with GDFs via interband transition. The results obtained well suggested that these impactful results were due to the light amplification of stimulated emission of THz radiation promoted by the GDP instabilities. Active controlling the PT symmetry of the GDPs is a new impactful technological trend to be able to implement the ultrafast direct gain switching functionality to the graphene plasmonic THz laser transistors. Furthermore, the author's group discovered a new mechanism of instability in GDPs called Coulomb-drag instability. The theoretical analyses suggest that the amplification gain could exceed the level of other existing mechanisms of instability. At this moment, however, the new mechanisms of the last two topics have yet to be experimentally verified, which is expected to be disclosed in the near future.

Acknowledgements

The author thanks Victor Ryzhii, Akira Satou, Stephane A. Boubanga-Tombet, Takayuki Watanabe, Deepika Yadav, Gen Tamamushi, Kenta Sugawara, Junki Mitsushio, Tetsuya Suemitsu, Hirokazu Fukidome, Maki Suemitsu, Youssef Tobah, Maxim Ryzhii, Vladimir Mitin, Michael S. Shur, Alexander A. Dubinov, Vyacheslav V. Popov, Vladimir Ya Aleshkin, Dmitry Svintsov, Wojciech Knap, Dmytro B. But, Ilya V. Gorbenko, Valentin Kachorovskii, Juan A. Delgado-Notario, and Yahya M. Meziani for their contributions. He also thanks Michel Dyakonov for his valuable discussion and encouragement.

The devices described in this chapter were fabricated at the Nanoelectronics and Spintronics Laboratory, RIEC, Tohoku University, Japan.

The work done by the author was supported by JSPS KAKENHI #23000008, 16H06361, 20K20349, and No. 21H04546, Japan, JSPS-RFBR Bilateral Joint-Research Program #120204801, and the Commissioned Research by NICT Grant No. 01301, Japan.

Conflict of interest

The authors declare no conflict of interest.

Author details

Taiichi Otsuji
Tohoku University, Sendai, Japan

*Address all correspondence to: taiichi.otsuji.e8@tohoku.ac.jp

IntechOpen

© 2023 The Author(s). Licensee IntechOpen. This chapter is distributed under the terms of the Creative Commons Attribution License (http://creativecommons.org/licenses/by/3.0), which permits unrestricted use, distribution, and reproduction in any medium, provided the original work is properly cited.

References

[1] Sengupta K, Nagatsuma T, Mittleman DM. Terahertz integrated electronic and hybrid electronic–photonic systems. Nature Electronics. 2018;**1**: 622-635. DOI: 10.1038/s41928-018-0173-2

[2] Tonouchi M. Cutting-edge terahertz technology. Nature Photonics. 2007;**1**: 97-105. DOI: 10.1038/s41928-018-0173-2

[3] Tredicucci A, Vitiello MS. Device concepts for graphene-based terahertz photnics. IEEE Journal of Selected Topics of Quantum Electronics. 2014;**20**: 8500109. DOI: 10.1109/JSTQE. 2013.2271692

[4] Novoselov KS, Geim AK, Morozov SV, Jiang D, Zhang Y, Dubonos SV, et al. Electric field effect in atomically thin carbon films. Science. 2004;**306**:666-669. DOI: 10.1126/ science.1102896

[5] Geim AK, Novoselov KS. The rise of graphene. Nature Materials. 2007;**26**: 183-191. DOI: 10.1038/nmat1849

[6] Ando T, Fowler AB, Stern F. Electronic properties of two-dimensional systems. Reviews of Modern Physics. 1982;**54**:437-672. DOI: 10.1103/ RevModPhys.54.437

[7] Ryzhii V, Satou A, Otsuji T. Plasma waves in two-dimensional electron-hole system in gated graphene heterostructures. Journal of Applied Physics. 2007;**101**:024509. DOI: 10.1063/ 1.2326904

[8] Grigorenko AN, Polini M, Novoselov KS. Graphene plasmonics. Nature Photonics. 2012;**6**:749-758. DOI: 10.1038/NPHOTON.2012.262

[9] Ryzhii V, Ryzhii M, Otsuji T. Negative dynamic conductivity of graphene with optical pumping. Journal of Applied Physics. 2007;**83**:083114. DOI: 10.1063/1.2717566

[10] Ryzhii M, Ryzhii V. Injection and population inversion in electrically induced p-n junction in graphene with split gates. Japanese Journal of Applied Physics. 2007;**46**:L151-L153. DOI: 10.1143/JJAP.46.L151

[11] Otsuji T, Boubanga-Tombet SA, Satou A, Ryzhii M, Ryzhii V. Terahertz-wave generation using graphene: Toward new types of terahertz lasers. IEEE Journal of Selected Topics in Quantum Electronics. 2013;**19**:8400209. DOI: 10.1109/JSTQE.2012.2208734

[12] Yadav D, Tamamushi G, Watanabe T, Mitsuhio J, Tobah Y, Sugawara K, et al. Terahertz light-emitting graphene-channel transistor toward single-mode lasing. Nanophotonics. 2018;7:741-752. DOI: 10.1515/nanoph-2017-0106

[13] Boubanga-Tombet S, Knap W, Yadav D, Satou A, But DB, Popov VV, et al. Room temperature amplification of terahertz radiation by grating-gate graphene structures. Physical Review X. 2020;**10**:031004. DOI: 10.1103/ PhysRevX.10.031004

[14] Boubanga-Tombet AS, Satou A, Yadav D, But DB, Knap W, Popov VV, et al. Paving the way for tunable graphene plasmonic THz amplifiers. Frontiers in Physics. 2021;**9**:726806. DOI: 10.3389/fphy.2021.726806

[15] Dyakonov M, Shur M. Shallow water analogy for a ballistic field effect transistor: New mechanism of plasma wave generation by dc current. Physical Review Letters. 1993;**71**:2465. DOI: 10.1103/PhysRevLett.71.2465

[16] Ryzhii V, Satou A, Shur MS. Transit time mechanism of plasma instability in high electron mobility transistors. Physica Status Solidi A: Applications and Materials Science. 2005;**202**:R113. DOI: 10.1002/pssa.200521018

[17] Aizin GR, Mikalopas J, Shur M. Current-driven plasmonic boom instability in three-dimensional gated periodic ballistic nanostructures. Physical Review B. 2016;**93**:195315. DOI: 10.1103/PhysRevB.93.195315

[18] Mikhailov SA. Plasma instability and amplification of electromagnetic waves in low-dimensional electron systems. Physical Review B. 1998;**58**:1517. DOI: 10.1103/PhysRevB.58.1517

[19] Sano E. Monte Carlo simulation of ultrafast electron relaxation in graphene. Applied Physics Express. 2011;**4**:085101. DOI: 10.1143/APEX.4.085101

[20] Foster MS, Aleiner IL. Slow imbalance relaxation and thermoelectric transport in graphene. Physical Review B. 2009;**79**:085415. DOI: 10.1103/PhysRevB.79.085415

[21] Winzer T, Malic E. Impact of Auger processes on carrier dynamics in graphene. Physical Review B. 2012;**85**: 2410404(R). DOI: 10.1103/PhysRevB.85.241404

[22] Kim R, Perebeinos V, Avouris P. Relaxation of optically excited carriers in graphene. Physical Review B. 2011;**84**: 075449. DOI: 10.1103/PhysRevB.84.075449

[23] Someya T, Fukidome H, Watanabe H, Yamamoto T, Okada M, Sozuki H, et al. Suppression of supercollision carrier cooling in high mobility graphene on SiC(000¯1). Physical Review B. 2017; **95**:165303. DOI: 10.1103/PhysRevB.95.165303

[24] Alymov G, Vyurkov V, Ryzhii V, Satou A, Svintsov D. Auger recombination in Dirac materials: A tangle of many-body effects. Physical Review B. 2018;**97**:205411. DOI: 10.1103/PhysRevB.97.205411

[25] Dubinov AA, Aleshkin VY, Morozov SV, Ryzhii V, Otsuji T. Terahertz plasmon-emitting graphene-channel transistor. Opto-Electronics Review. 2019;**27**:345-347. DOI: 10.1016/j.opelre.2019.11.003

[26] Dubinov AA, Alehkin YV, Mitin V, Otsuji T, Ryzhii V. Terahertz surface plasmons in optically pumped graphene structures. Journal of Physics: Condensed Matter. 2011;**23**:145302. DOI: 10.1088/0953-8984/23/14/145302

[27] Watanabe T, Fukushima T, Yabe Y, Boubanga Tombet SA, Satou A, Dubinov AA, et al. The gain enhancement effect of surface plasmon polaritons on terahertz stimulated emission in optically pumped monolayer graphene. New Journal of Physics. 2013;**15**:075003. DOI: 10.1088/1367-2630/15/7/075003

[28] Ryzhii V, Otsuji T, Shur MS. Graphene based plasma-wave devices for terahertz applications. Applied Physics Letters. 2020;**116**:140501. DOI: 10.1063/1.5140712

[29] Popov VV, Fateev DV, Otsuji T, Meziani YM, Coquillat D, Knap W. Plasmonic terahertz detection by a double-grating-gate field-effect transistor structure with an asymmetric unit cell. Applied Physics Letters. 2011; **99**:243504. DOI: 10.1063/1.3670321

[30] Koseki K, Ryzhii V, Otsuji T, Popov VV, Satou A. Giant plasmon instability in a dual-grating-gate graphene field-effect transistor. Physical Review B. 2016;**93**:245408. DOI: 10.1103/PhysRevB.93.245408

[31] Ryzhii V, Ryzhii M, Mitin V, Shur MS, Otsuji T. S-shaped current-voltage characteristics of n+−i-n-n+ graphene field-effect transistors due the coulomb drag of quasi-equilibrium electrons by ballistic electrons. Physical Review Applied. 2021;**16**:014001. DOI: DOI

[32] Ryzhii V, Ryzhii M, Satou A, Mitin V, Shur MS, Otsuji T. Ballistic injection terahertz plasma instability in graphene n + −i-n-n + field-effect transistors and lateral diodes. Physica Status Solidi A. 2022;**219**:2100694. DOI: 10.1002/pssa.202100694

[33] Ryzhii V, Ryzhii M, Mitin V, Shur MS, Otsuji T. Coulomb electron drag mechanism of terahertz plasma instability in n+−i-n-n+ graphene FETs with ballistic injection. Applied Physics Letters. 2021;**119**:093501. DOI: 10.1063/5.0061722

[34] Ryzhii V, Ryzhii M, Satou A, Otsuji T, Mitin V, Shur MS. Effect of coulomb carrier drag and terahertz plasma instability in p+−p-i-n-n+ graphene tunneling transistor structures. Physical Review Applied. 2021;**16**:064054. DOI: 10.1103/PhysRevApplied.16.064054

[35] Mayorov AS, Gorbachev RV, Morozov SV, Britnell L, Jalil R, Ponomarenko LA, et al. Micrometer-scale ballistic transport in encapsulated graphene at room temperature. Nano Letters. 2011;**11**:2396-2399. DOI: 10.1021/nl200758b

[36] Bandurin DA, Torre I, Krishna Kumar R, Ben Shalom M, Tomadin A, Principi A, et al. Negative local resistance caused by viscous electron backflow in graphene. Science. 2016;**351**:1055-1058. DOI: 10.1126/science.aad0201

[37] Jena D. Tunneling transistors based on graphene and 2-D crystals. Proceedings of the IEEE. 2013;**101**: 1585-1602. DOI: 10.1109/JPROC.2013.2253435

[38] Gorbachev RV, Geim AK, Katsunelson MI, Novoselov KS, Tudorovskiy T, Grigorieva IV, et al. Strong coulomb drag and broken symmetry in double-layer graphene. Nature Physics. 2012;**8**:896-901. DOI: 10.1038/nphys2441

[39] Gribnikov ZS, Vagidov NZ, Mitin VV. Tow-stream instability and oscillatory regimes induced in ballistic diodes and field-effect transistors. Journal of Applied Physics. 2000;**88**: 6736-6745. DOI: 10.1063/1.1322383

[40] Bender CM, Boettcher S. Real spectra in non-Hermitian Hamiltonians having PT symmetry. Physical Review Letters. 1998;**80**:5243-5246. DOI: 10.1103/PhysRevLett.80.5243

[41] Ramezani H, Kottos T. Unidirectional nonlinear PT-symmetric optical structures. Physical Review A. 2010;**82**:04383. DOI: 10.1103/PhysRevA.82.043803

[42] Endoh N, Akiyama S, Tashima K, Suwa K, Kamogawa T, Kohama R, et al. High-quality few-layer graphene on single-crystalline SiC thin film grown on affordable wafer for device applications. Nanomaterials. 2021;**11**:392. DOI: 10.3390/nano11020392

[43] Fukidome H, Kawai Y, Fromm F, Kotsugi M, Handa H, Ide T, et al. Suemitsu M: Precise control of epitaxy of graphene by microfabricating SiC substrate. Applied Physics Letters. 2012; **101**:041605. DOI: 10.1063/1.4740271

Section 3

Biology and Terahertz Radiation

Chapter 6

Introduction to the Biological Effects of Terahertz Radiation

Robin-Cristian Bucur-Portase

Abstract

Terahertz (THz) radiation has been noted to affect biological organisms to a unique degree with various effects ranging from modifications brought to protein activity to epigenetic changes that lead to altered metabolism or reproduction. These effects are classified into thermal and non-thermal, with the former being caused by THz's capacity to induce localised thermal changes while the latter involves more complex interactions with cells' macromolecules which are poorly understood. Terahertz's ability to enhance actin polymerisation and alter gene expression leads to a number of possible applications in agriculture, as it has been observed that certain plant species have higher growth speeds post-exposure, and medicine, with cancer's rapid division being possibly slowed down.

Keywords: terahertz, histology, cytology, genetics, medicine, agriculture

1. Introduction

Terahertz radiation (T-rays or THz) is known for inducing various changes in biological macromolecules and structures despite being a form of non-ionising radiation [1]. Genetic changes, metabolic shifts and general alterations in cells', tissues' and whole organisms' functioning and structure have been recorded [2–5]. Research is still in its early stages surrounding this type of radiation's effects and as such its documented effects are not very well understood.

THz is known for inducing thermal and non-thermal changes in biological tissues. In most cases, it is unclear which of the types of alterations occur or whether it is a combination of the two that gives rise to the observed phenomena. In some instances, the entirety of the energy provided by THz gets transformed into thermal energy.

An important factor surrounding terahertz's effects is the presence of water in the organism exposed. Its absorption coefficient is 300 cm^{-1} at 1.5 THz which makes it a powerful T-rays absorption material [6]. Thus, in some instances, water can act as a shield or as an agent that exclusively stops non-thermal effects but enhances the latter form of energy transfer, all being reliant on the proportional quantity of water present in the tissue analysed.

To account for the thermal effects of THz, a heat transfer model can be developed. For preliminary characterisation of its effects on biological materials the relationships that describe the rate of heat generation and rise in temperature suffice.

The following protocol has been adapted from multiple sources. See [7–10].

Due to the air-skin interface representing an optical system of two very different mediums, the refracted beam's angle can be found by applying Snell's law:

$$n_1 sin\,\theta_1 = n_2 sin\,\theta_2 \tag{1}$$

where n_1 and n_2 are the refractive indexes of air and skin, respectively, and θ_1 and θ_2 are the incident and refracted angles, respectively.

Next, the Fresnel equation helps determine how much THz power is lost to reflection:

$$R_S = \frac{1}{2}\left(\frac{tan^2(\theta_1 - \theta_2)}{tan^2(\theta_1 + \theta_2)} + \frac{sin^2(\theta_1 - \theta_2)}{sin^2(\theta_1 + \theta_2)}\right) \tag{2}$$

where R_S is the THz power lost due to reflection or specular reflectance.

It should be noted that R_S increases exponentially for angles of incidence greater than 60 degrees.

After accounting for this reflective loss, the final spectral flux that passes through the tissue is provided by Lambert–Beer's law:

$$A = \frac{\Phi_{e,\lambda,f}}{\Phi_{e,\lambda,o}} = e^{(-\mu_a d)} \tag{3}$$

where A is absorbance, $\Phi_{e,\lambda,f}$ is the final spectral flux, $\Phi_{e,\lambda,o}$ is the initial spectral flux of the beam that hits the skin, μ_a is the absorption coefficient, and d is the distance traversed by the radiation. It is worth noting that a new set of equations is to be applied each time the medium changes to account for the different refractive index (n) and tissue absorption coefficient (μ_a). Moreover, the Poynting vector's relationships may also be of use in calculating the value of the final spectral flux ($\Phi_{e,\lambda,f}$):

$$\Phi_{e,f} \approx \int_\Sigma \langle|S|\rangle\, cos\,\theta_1 d\Sigma \tag{4}$$

where $\Phi_{e,f}$ is the final radiant flux, $\langle|S|\rangle$ is the time average of the Poynting vector, θ_1 is the incident angle of the beam, and A is the area of tissue exposed.

The final spectral flux is then obtained by applying the relationship:

$$\Phi_{e,\lambda,f} = \frac{\partial\Phi_{e,f}}{\partial\lambda} \tag{5}$$

The equations surrounding the rise of temperature in biological tissues are governed by the following formula:

$$R(x,y,z) = \mu_a(x,y,z)\Phi_{e,\lambda,f}(x,y,z) \tag{6}$$

where R is the rate of heat generation at a point of Cartesian coordinates x, y, z in the tissue measured in Wnm^{-1}, μ_a is the local absorption coefficient, and $\Phi_{e,\lambda,f}$ is the radiometric spectral flux. After finding the value of R, the equation that describes the increase in temperature can be used:

$$\Delta T(x,y,z) = \frac{R(x,y,z)\Delta t}{\rho c} \tag{7}$$

where ΔT is the localised temperature rise, Δt is the duration of the exposure, ρ is the tissue density, and c is the tissue's specific heat capacity.

Eqs. (6) and (7) describe optical phenomena taking place for one point of Cartesian coordinates x, y and z. To adapt them for a tissue area, the average THz absorption coefficient of the tissue and the spectral irradiance need to be calculated. The latter is obtained using the following relationship:

$$E_{e,\lambda,f} = \frac{\partial \Phi_{e,\lambda,f}}{\partial A} \tag{8}$$

where A is the area exposed.

For more advanced modelling of the heat transfer of THz in tissues, the Pennes' bioheat equation and its Cartesian transformation can be used in modern computational and simulation techniques. Discussing these, however, is beyond the scope of this chapter.

Moreover, taking into account the complex interaction between water and terahertz radiation, a complete model describing heat generation would require taking into account rotational transitions, dielectric properties of water and the coupled oscillating electric and magnetic fields' inter-relational excitations through a system of several equations such as Debye's model, Maxwell's set, Newton's cooling and Von Neumann's. Multiple accepted models may arise just as in the case of the mathematical relationships used to describe the phenomena that occur with microwave heating [11].

2. Interaction with animal tissue and cells

To properly interact with mammalian organisms' tissues, terahertz radiation must first bypass the skin. The degree to which T-rays penetrate through the epidermis is not fully known due to the under-characterised optical properties of melanin, sulphated proteoglycans and other major components of the epidermis. The dermis, unlike its superficial homologue, has a composition that is very rich in glycosaminoglycans that are capable of holding large amounts of water molecules which impedes the passage of terahertz radiation. This makes it behave very similarly to free water from a reflectivity point of view [12]. Moreover, different concentrations and types of melanin drastically influence the optical properties of the dermis [13, 14].

There are conflicting reports regarding the importance that collagen has regarding skin THz absorption. In one study, it was found that isolated skin collagen has an insignificant response to T-rays [12]. In another, the authors found that the heat denaturation of collagen fibres is directly related to a decrease in the absorption capabilities of the dermis [15].

While THz can interact wildly differently with different organisms, there are a couple of effects that occur irrespective of the structure exposed. These include the formation of reactive oxygen species (ROS), reactive nitrogen species (RNS), cellular, nuclear and lysosomal membrane destabilisation and general hyperthermic shock responses [16–21].

T-rays also affect the structure and activity of various proteins and enzymes [22].

Moreover, terahertz radiation has been placed under the spotlight due to its observed effects on genetic material. Specifically, it manages to break the hydrogen bonds that hold double-stranded DNA together by creating interference within its double helix structure through presumed nonthermal effects [23]. This type of EM radiation is also capable of demethylating various genes with a wide range of effects, showing promise as a potential cancer treatment [24]. It has also been shown capable of inducing histone H2AX phosphorylation in human skin fibroblasts, effects which last for at least 24 h [25]. THz has even been reported to induce aneuploidy in human lymphocytes [26] and of changing the gene expression of mouse stem cells so that they more readily differentiate into adipocytes via the activation of the transcription factor peroxisome proliferator-activated receptor gamma (PPARG) [27].

Additionally, terahertz radiation is capable of interfering with intracellular actin structures. There exist conflicting reports with some authors suggesting that THz collapses actin filaments [28] while others claim an enhancement of actin fibrils lengthening dynamics [29]. It has also been observed that cellular division is inhibited via actin-mediated interferences [30].

3. Medical implications

3.1 THz as a data acquisition tool

Terahertz radiation has seen applications in laboratory environments, helping detect the presence and concentration of various biomolecules with fine precision. The main method used is spectroscopy which has various subsets such as vibrational and attenuated total reflection [31–36]. Considerable advances have been made with regard to THz analysis techniques of biomolecules in aqueous environments [37].

THz spectroscopy is very precise, being able to distinguish between different enantiomers and going as far as identifying the different hydrogen isotopes present within the structures of amino acids [38]. It is also able to assess the configuration and hydrogen bonding of various molecules [39]. Finally, by coupling T-rays to various subwavelength probes, a resolution that reaches ångströms can be achieved [40, 41].

There are also numerous claimed applications for terahertz radiation in *in vivo* and *ex vivo* medical analyses such as biopsies and dental scans [42, 43]. Various tissues interact differently with THz. Skin and adipose tissue have similar absorption coefficients but both differ from muscles. The main determining factor seems to be water content [44]. Thus, terahertz scanning methods should be able to distinguish between different tissues or even cell types based on their hydration, complementing existing medical screening techniques.

Several materials do not interact strongly with THz and could be used in diagnostic and laboratory scans. Such an example is high-density polyethylene (HDPE) (**Figure 1**).

Acrylic is not very suitable due to its peak absorbance coefficient reaching values of 15 cm^{-1}. Cycloolefins seem to be the optimal window material choice, as they display close to no optical interaction with THz. Other materials such as teflon have very low absorption spectra but should not be used due to their environmental and health impact [45].

3.2 Curative and toxic properties

Terahertz radiation has also seen therapeutic usage. It can demethylate cancer cells and increase the speed of regeneration at injuries' site [46–48]. It has even been found to be beneficial against psoriasis [49] and in the recovery from an acute ischemic stroke [50].

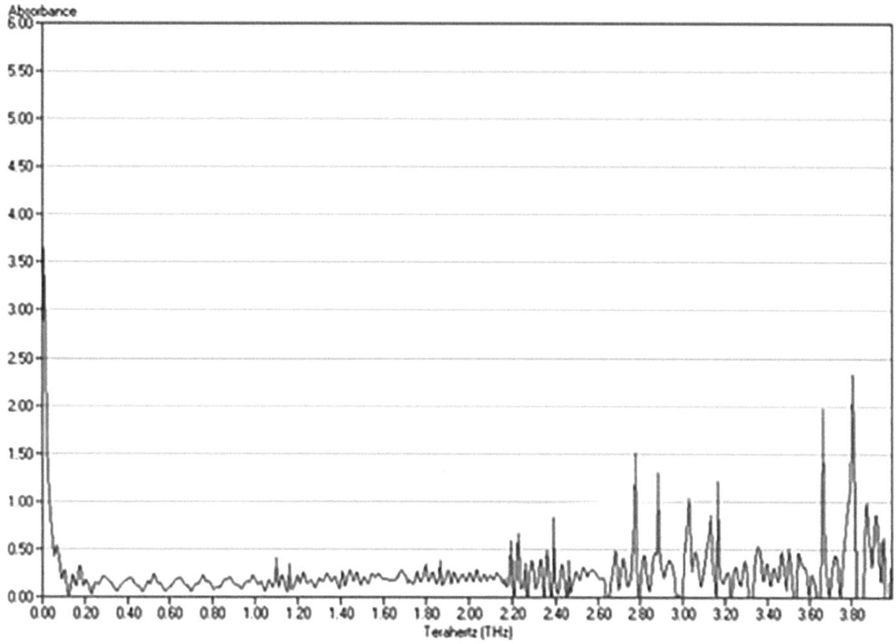

Figure 1.
THz absorption spectrum of HDPE. Image owned by the author. See notes for more information.

However, the use of these therapeutic effects faces barriers due to the adverse effects often caused by terahertz radiation exposure. Most reports describe general inflammation responses and apoptosis that have been recorded across a vast array of cell types and species. Loss of adhesion to basal membrane, cellular permeability increases, lysis and marked increases in cell growth factors and cytokines have all been connected to T-rays exposure, although these seem to mainly stem from its thermal effects [51–53]. However, there are also reports that contrast these findings, albeit very few [54, 55].

Nervous tissue cells in particular have been found to release their intracellular proteolytic enzymes and suffer membrane protein changes under THz. No great morphological changes such as axon number or size have been recorded [56, 57]. Various types of neurons also show different reactions to T-rays as seen by their marked changes in neurotransmitter production [58]. There also seems to be a threshold of 0.15 THz under which exposure will not induce any detectable adverse reactions [59].

3.3 Other uses

It has been reported that terahertz radiation can be successfully used to sterilise liquids [60].

4. Effects on plant and fungi matter

4.1 Plants

There are very few studies that investigate the use of terahertz radiation on plants. The few that do exist have mostly shown variable effects. Depending on the dosage, there has been an increase, decrease or no change in the growth rate of adult plants [61]. Meanwhile, a report shows that dry wheat seeds exposed during their dormant state and with no additional T-rays received afterwards have a lowered rate of successful germination contrasted by variable growth speed changes [62].

The absorption coefficients of starch, chlorophyll and other plant compound have been determined [63–67]. Moreover, the separation of seeds based on their bacterial blight resistance or genetic modifications has been described [68, 69].

4.2 Fungi

There is almost no literature on the effects of THz exposure on fungi with the exception of one report that found an increase in cellular growth in the *S. cerevisiae* yeast [70].

5. Environmental effects

Terahertz radiation may be able to penetrate the soil to a considerable depth and, having in mind its effects on tissues of animals and plants alike, disturb local ecological systems. This raises concerns regarding the rising trend of new agricultural tools that implement this type of EM radiation as a means to keep track of plant development or soil content in microplastics [71–76]. Locations suffering from a generally dry climate or desertification are at the highest risk of being affected by THz due to the lack of water that could otherwise act as a barrier.

6. Conclusion

The general applications and risks of terahertz radiation usage on living matter prove this type of EM radiation particularly interesting. With applications including cancer treatment and detection, enhanced agricultural output, increased tissue healing and new genetic engineering tools, the technology may lead to an improvement in most scientific fields surrounding biological organisms. However, the lack of understanding of how THz exhibits its effects on living matter remains this field's weakest point.

The most important steps that should follow are the precise determination of the health effects of exposure to terahertz radiation, taking into account the induced genetic, metabolic and structural changes surrounding both short- and long-term exposure. Characterisation of thermal and non-thermal effects would be the following step to better understand the full complexity surrounding its mechanism of action. Moreover, any experiments performed should outline all the important variables pertaining to them in a standardised fashion. These factors must include the frequency, intensity, source of terahertz, whether the samples are exposed to continuous wave (CW) or pulsating lasers, any temperature changes and the water content. This list is not exhaustive.

Acknowledgements

The author is grateful for the financial support provided by the University of Medicine and Pharmacy “Carol Davila”.

Conflict of interest

The author declares no conflict of interest.

Notes

The spectrometer TeraView (TPS Spectra 3000) was used for the production of **Figure 1**. It is capable of emitting terahertz radiation through an optical tunable Ti: Sapphire ultrashort pulsed laser with a spectral range of 0.1–3 THz. During the identification of the absorption spectrum of HDPE, it was set to 150 mW, half its original power.

Thanks

A special note of thanks is addressed to my brother, family, friends and tutors who are tirelessly supportive of all of my endeavours.

Author details

Robin-Cristian Bucur-Portase
University of Medicine and Pharmacy Carol Davila, Bucharest, Romania

*Address all correspondence to: robinbucur.publish@gmail.com

IntechOpen

© 2023 The Author(s). Licensee IntechOpen. This chapter is distributed under the terms of the Creative Commons Attribution License (http://creativecommons.org/licenses/by/3.0), which permits unrestricted use, distribution, and reproduction in any medium, provided the original work is properly cited.

References

[1] Yakasai IK, PgE A, Begum F. Review of porous core photonic crystal fibers for terahertz waveguiding. Optik. 2021;**229**: 166284

[2] Cheon H, Paik JH, Choi M, Yang HJ, Son JH. Detection and manipulation of methylation in blood cancer DNA using terahertz radiation. Scientific Reports. 2019;**9**(1):6413

[3] Soloveva AG, Peretyagin PV, Polyakova AG, Didenko NV. The influence of terahertz radiation on biochemical metabolism of blood in the experiment. The European Physical Journal Conferences. 2018;**195**:10013

[4] Inhibition of cancer cell migration and glycolysis by terahertz wave modulation via altered chromatin accessibility. Research. 2022;**2022**: 9860679. DOI: 10.34133/2022/9860679

[5] Hu E, Zhang Q, Shang S, Jiang Y, Lu X. Continuous wave irradiation at 0.1 terahertz facilitates transmembrane transport of small molecules. iScience. 2022;**25**(3):103966

[6] Kindt JT, Schmuttenmaer CA. Far-infrared dielectric properties of polar liquids probed by femtosecond terahertz pulse spectroscopy. The Journal of Physical Chemistry. 1996;**100**(24): 10373-10379

[7] Wilmink GJ, Grundt JE. Invited review article: Current state of research on biological effects of terahertz radiation. Journal of Infrared, Millimeter, and Terahertz Waves. 2011; **32**(10):1074-1122

[8] McCluney WR. Introduction to Radiometry and Photometry. 1st ed. London: Artech House; 1994. p. 417

[9] Griffiths DJ. Introduction to Electrodynamics. 3rd ed. Upper Saddle River, N.J: Prentice Hall; 1999. p. 576

[10] 14:00–17:00. ISO 9288:2022. ISO. [cited 2023 Jan 21]. [Internet]. Available from: https://www.iso.org/standard/82088.html

[11] Yang B, Huang H, Zhou L, Jin H. Method for solving the microwave heating temperature distribution of the TE10 mode. PRO. 2022;**10**(7):1377

[12] Tewari P, Culjat MO, Taylor ZD, Suen JY, Burt BO, Lee H, et al. Role of collagen in terahertz absorption in skin. In: Advanced Biomedical and Clinical Diagnostic Systems VII. SPIE. 2009. 71691A. pp. 245-252. Available from: https://www.spiedigitallibrary.org/conference-proceedings-of-spie/7169/71691A/Role-of-collagen-in-terahertz-absorption-in-skin/10.1117/12.816433.full

[13] Nikitkina AI, Bikmulina P, Gafarova ER, Kosheleva NV, Efremov YM, Bezrukov EA, et al. Terahertz radiation and the skin: A review. Journal of Biomedical Optics. 2021;**26**(4):043005

[14] Peralta XG, Lipscomb D, Wilmink GJ, Echchgadda I. Terahertz spectroscopy of human skin tissue models with different melanin content. Biomedical Optics Express. 2019;**10**(6): 2942-2955

[15] Mizuno M, Yaekashiwa N, Watanabe S. Analysis of dermal composite conditions using collagen absorption characteristics in the THz range. Biomedical Optics Express. 2018; **9**(5):2277-2283

[16] Hornemann A, Eichert DM, Hoehl A, Tiersch B, Ulm G,

Ryadnov MG, et al. Investigating membrane-mediated antimicrobial peptide interactions with synchrotron radiation far-infrared spectroscopy. Chemphyschem : A European Journal of Chemical Physics and Physical Chemistry. 2022;**23**(4):e202100815

[17] Zou Y, Liu Q, Yang X, Huang HC, Li J, Du LH, et al. Label-free monitoring of cell death induced by oxidative stress in living human cells using terahertz ATR spectroscopy. Biomedical Optics Express. 2017;**9**(1):14-24

[18] Ramzan R, Staniek K, Kadenbach B, Vogt S. Mitochondrial respiration and membrane potential are regulated by the allosteric ATP-inhibition of cytochrome c oxidase. Biochimica et Biophysica Acta. 2010;**1797**(9):1672-1680

[19] Kültz D. Evolution of the cellular stress proteome: From monophyletic origin to ubiquitous function. The Journal of Experimental Biology. 2003; **206**(Pt 18):3119-3124

[20] Wilmink GJ, Roth CL, Ibey BL, Ketchum N, Bernhard J, Cerna CZ, et al. Identification of microRNAs associated with hyperthermia-induced cellular stress response. Cell Stress & Chaperones. 2010;**15**(6):1027-1038

[21] Demidova E, Goryachkovskaya T, Mescheryakova I, Malup T, Popik B, Vinokurov N, et al. Impact of terahertz radiation on stress-sensitive genes of *E. Coli* cell. IEEE Transactions on Terahertz Science and Technology. 2016; **6**:1-7

[22] Govorun VM, Tretiakov VE, Tulyakov NN, Fleurov VB, Demin AI, Volkov AY, et al. Far-infrared radiation effect on the structure and properties of proteins. International Journal of Infrared and Millimeter Waves. 1991; **12**(12):1469-1474

[23] Chitanvis SM. Can low-power electromagnetic radiation disrupt hydrogen bonds in dsDNA? Journal of Polymer Science Part B: Polymer Physics. 2006;**44**:2740-2747. Available from: https://onlinelibrary.wiley.com/doi/abs/10.1002/polb.20910

[24] Cheon H, Yang HJ, Choi M, Son JH. Effective demethylation of melanoma cells using terahertz radiation. Biomedical Optics Express. 2019;**10**(10): 4931-4941

[25] Effects of high intensity non-ionizing terahertz radiation on human skin fibroblasts - PMC [Internet]. [cited 2023 Jan 22]. Available from: https://www.ncbi.nlm.nih.gov/pmc/articles/PMC8606137/

[26] Korenstein-Ilan A, Barbul A, Hasin P, Eliran A, Gover A, Korenstein R. Terahertz radiation increases genomic instability in human lymphocytes. Radiation Research. 2008; **170**(2):224-234

[27] Bock J, Fukuyo Y, Kang S, Phipps ML, Alexandrov LB, Rasmussen KØ, et al. Mammalian stem cells reprogramming in response to terahertz radiation. PLoS One. 2010;**5**(12):e15806

[28] Yamazaki S, Harata M, Ueno Y, Tsubouchi M, Konagaya K, Ogawa Y, et al. Propagation of THz irradiation energy through aqueous layers: Demolition of actin filaments in living cells. Scientific Reports. 2020;**10**(1): 9008

[29] Yamazaki S, Harata M, Idehara T, Konagaya K, Yokoyama G, Hoshina H, et al. Actin polymerization is activated by terahertz irradiation. Scientific Reports. 2018;**8**(1):9990

[30] Yamazaki S, Ueno Y, Hosoki R, Saito T, Idehara T, Yamaguchi Y, et al.

THz irradiation inhibits cell division by affecting actin dynamics. PLoS One. 2021;**16**(8):e0248381

[31] Peng Y, Shi C, Zhu Y, Gu M, Zhuang S. Terahertz spectroscopy in biomedical field: A review on signal-to-noise ratio improvement. PhotoniX. 2020;**1**(1):12

[32] Wagh MP, Sonawane YH, Joshi OU. Terahertz technology: A boon to tablet analysis. Indian Journal of Pharmaceutical Sciences. 2009;**71**(3): 235-241

[33] Wei L, Yu L, Jiaoqi H, Guorong H, Yang Z, Weiling F. Application of terahertz spectroscopy in biomolecule detection. Frontiers in Laboratory Medicine. 2018;**2**(4):127-133

[34] Jeong SY, Cheon H, Lee D, Son JH. Determining terahertz resonant peaks of biomolecules in aqueous environment. Optics Express. 2020;**28**(3):3854-3863

[35] El Haddad J, Bousquet B, Canioni L, Mounaix P. Review in terahertz spectral analysis. TrAC Trends in Analytical Chemistry. 2013;**44**:98-105

[36] Mancini T, Mosetti R, Marcelli A, Petrarca M, Lupi S, D'Arco A. Terahertz spectroscopic analysis in protein dynamics: Current status. Radiation. 2022;**2**(1):100-123

[37] Weisenstein C, Wigger AK, Richter M, Sczech R, Bosserhoff AK, Bolívar PH. THz detection of biomolecules in aqueous environments—Status and perspectives for analysis under physiological conditions and clinical use. Journal of Infrared, Millimeter, and Terahertz Waves. 2021;**42**(6):607-646

[38] Hufnagle DC, Taulbee-Combs AR, Spendel WU, Pacey GE. Collective mode frequency shifts in l-serine and a series of isotopologues in the terahertz regime. Talanta. 2012;**88**:61-65

[39] Walther M, Plochocka P, Fischer B, Helm H, Uhd JP. Collective vibrational modes in biological molecules investigated by terahertz time-domain spectroscopy. Biopolymers. 2002;**67**(4–5):310-313

[40] Cocker TL, Jelic V, Hillenbrand R, Hegmann FA. Nanoscale terahertz scanning probe microscopy. Nature Photonics. 2021;**15**(8):558-569

[41] Wächter M, Nagel M, Kurz H. Tapered photoconductive terahertz field probe tip with subwavelength spatial resolution. Applied Physics Letters. 2009;**95**(4):041112

[42] Terasense. Medical Diagnostics [Internet]. Available from: https://terasense.com/applications/medical-diagnostics/ [Accessed: 2023-01-22]

[43] Sun Q, He Y, Liu K, Fan S, Parrott EPJ, Pickwell-MacPherson E. Recent advances in terahertz technology for biomedical applications. Quantitative Imaging in Medicine and Surgery. 2017; **7**(3):345-355

[44] Wilmink GJ, Ibey BL, Tongue T, Schulkin B, Laman N, Peralta XG, et al. Development of a compact terahertz time-domain spectrometer for the measurement of the optical properties of biological tissues. Journal of Biomedical Optics. 2011;**16**(4):047006

[45] Withayachumnankul W, Png GM, Yin X, Atakaramians S, Jones I, Lin H, et al. T-Ray sensing and imaging. Proceedings of the IEEE. 2007;**95**(8): 1528-1558

[46] Son JH, Cheon H. Toward cancer treatment using terahertz radiation:

Demethylation of cancer cells. In: Next-Generation Spectroscopic Technologies XIII. SPIE; 2020. p. 1139002 Available from: https://www.spiedigitallibrary.org/conference-proceedings-of-spie/11390/1139002/Toward-cancer-treatment-using-terahertz-radiation–demethylation-of-cancer/10.1117/12.2557655.full

[47] Lukin SY, Soldatov YP, Stogov MV. The complex correction of the pathophysiological disorders in the patients presenting with orthopedic and traumatological problems with the application of the electromagnetic waves of the terahertz range at the radiation frequencies of nitric oxide. Voprosy Kurortologii, Fizioterapii, i Lechebnoĭ Fizicheskoĭ Kultury. 2018;**95**(6):58-66

[48] Kozlov E, Soldatov YP, Stogov M, Shchurova E, Kireeva E. The results of the impact of electromagnetic waves of the terahertz range on the tissues of the elbow joint with the consequences of its injury. Genij Ortopedii. 2022;**28**:328-332

[49] Titova LV, Ayesheshim AK, Golubov A, Rodriguez-Juarez R, Woycicki R, Hegmann FA, et al. Intense THz pulses down-regulate genes associated with skin cancer and psoriasis: A new therapeutic avenue? Scientific Reports. 2013;**3**(1):2363

[50] Reukov A, Naymushin A, Simakov K, Moroshkin V, Kozlenok A, Presnukhina A. Use of the infrared radiation modulated by terahertz frequencies in complex therapy of patients with acute ischemic stroke. Arterial'naya Gipertenziya (Arterial Hypertension). 2016;**22**:94-102

[51] Cherkasova OP, Serdyukov DS, Ratushnyak AS, Nemova EF, Kozlov EN, Shidlovskii YV, et al. Effects of terahertz radiation on living cells: A review. Optics and Spectroscopy. 2020;**128**(6):855-866

[52] Cherkasova OP, Serdyukov DS, Nemova EF, Ratushnyak AS, Kucheryavenko AS, Dolganova IN, et al. Cellular effects of terahertz waves. Journal of Biomedical Optics. 2021;**26**(9):090902

[53] Hintzsche H, Stopper H. Effects of terahertz radiation on biological systems. Critical Reviews in Environmental Science and Technology. 2012;**42**(22):2408-2434

[54] Williams R, Schofield A, Holder G, Downes J, Edgar D, Harrison P, et al. The influence of high intensity terahertz radiation on mammalian cell adhesion, proliferation and differentiation. Physics in Medicine and Biology. 2013;**58**(2):373-391

[55] Zeni O, Gallerano GP, Perrotta A, Romanò M, Sannino A, Sarti M, et al. Cytogenetic observations in human peripheral blood leukocytes following in vitro exposure to THz radiation: A pilot study. Health Physics. 2007;**92**(4):349-357

[56] Zhang J, Li S, Le W. Advances of terahertz technology in neuroscience: Current status and a future perspective. iScience. 2021;**24**(12):103548

[57] Zhao L, Yi R, Liu S, Chi Y, Tan S, Dong J, et al. Biological responses to terahertz radiation with different power density in primary hippocampal neurons. PLOS ONE. 2023;**18**(1):e0267064

[58] Tan SZ, Tan PC, Luo LQ, Chi YL, Yang ZL, Zhao XL, et al. Exposure effects of terahertz waves on primary neurons and neuron-like cells under nonthermal conditions. Biomedical and Environmental Sciences. 2019;**32**(10):739-754

[59] Bourne N, Clothier RH, D'Arienzo M, Harrison P. The effects of

terahertz radiation on human keratinocyte primary cultures and neural cell cultures. Alternatives to Laboratory Animals. 2008;**36**(6):667-684

[60] Terahertz sterilization and disinfection liquid and preparation method thereof. CN106620753A. 2017 [Internet] [cited 2023 Jan 21]. Available from: https://patents.google.com/patent/CN106620753A/en

[61] Fedorov VI, Popova SS, Pisarchik AN. Dynamic effects of Submillimeter wave radiation on biological objects of various levels of organization. International Journal of Infrared and Millimeter Waves. 2003; **24**(8):1235-1254

[62] Bucur-Portase RC. The effects of terahertz radiation on the development of biological organisms I: Wheat seeds. In: Tiginyanu I, Sontea V, Railean S, editors. 5th International Conference on Nanotechnologies and Biomedical Engineering. Cham: Springer International Publishing; 2022. pp. 483-488 (IFMBE Proceedings)

[63] Coquillat D, O'Connor E, Brouillet EV, Meriguet Y, Bray C, Nelson DJ, et al. Investigating the low-frequency vibrations of chlorophyll derivatives using terahertz spectroscopy. In: Razeghi M, Baranov AN, editors. Terahertz Emitters, Receivers, and Applications XII [Internet]. San Diego, United States: SPIE; 2021 [cited 2023 Jan 22]. p. 13. Available from: https://www.spiedigitallibrary.org/conference-proceedings-of-spie/11827/2593589/Investigating-the-low-frequency-vibrations-of-chlorophyll-derivatives-using-terahertz/10.1117/12.2593589.full

[64] Qu Y, Zhang S, Lian Y, Kuang T. Function of terahertz spectra in monitoring the decomposing process of biological macromolecules and in investigating the causes of photoinhibition. Science China. Life Sciences. 2017;**60**(3):307-312

[65] Nakajima S, Horiuchi S, Ikehata A, Ogawa Y. Determination of starch crystallinity with the Fourier-transform terahertz spectrometer. Carbohydrate Polymers. 2021;**262**:117928

[66] Wang T, Wang S, Zhai C, Wang L, Xie Y, Li Q, et al. Study of starch aging characteristics based on terahertz technology. Food Science & Nutrition. 2021;**9**(8):4431-4439

[67] Nakajima S, Shiraga K, Suzuki T, Kondo N, Ogawa Y. Quantification of starch content in germinating mung bean seedlings by terahertz spectroscopy. Food Chemistry. 2019;**294**

[68] Zhang J, Yang Y, Feng X, Xu H, Chen J, He Y. Identification of bacterial blight resistant Rice seeds using terahertz imaging and hyperspectral imaging combined with convolutional neural network. Frontiers in Plant Science. 2020;**11**. Available from: https://www.frontiersin.org/articles/10.3389/fpls.2020.00821

[69] Qin B, Li Z, Chen T, Chen Y. Identification of genetically modified cotton seeds by terahertz spectroscopy with MPGA-SVM. Optik. 2017;**142**: 576-582

[70] Hadjiloucas S, Chahal MS, Bowen JW. Preliminary results on the non-thermal effects of 200-350 GHz radiation on the growth rate of *S. cerevisiae* cells in microcolonies. Physics in Medicine and Biology. 2002;**47**(21): 3831-3839

[71] Zahid A. Terahertz sensing at Nano-scale for future agriculture. Research and Reviews: Journal of Botanical Sciences. 2018;7:4

[72] Dworak V, Augustin S, Gebbers R. Application of terahertz radiation to soil measurements: Initial results. Sensors. 2011;**11**(10):9973-9988

[73] Zhao S, Zhang Y, Qiu Z, He Y, Zhang Y. Towards a fast and generalized microplastic quantification method in soil using terahertz spectroscopy. Science of The Total Environment. 2022; **841**:156624

[74] Zahid A, Abbas HT, Heidari H, Imran M, Alomainy A, Abbasi QH. Electromagnetic properties of plant leaves at terahertz frequencies for health status monitoring. In: 2019 IEEE MTT-S International Microwave Biomedical Conference (IMBioC) [Internet]. Nanjing, China: IEEE; 2019 [cited 2023 Jan 22]. pp. 1-4. Available from: https://ieeexplore.ieee.org/document/8777782/

[75] Born N, Behringer D, Liepelt S, Beyer S, Schwerdtfeger M, Ziegenhagen B, et al. Monitoring plant drought stress response using terahertz time-domain spectroscopy. Plant Physiology. 2014;**164**(4):1571-1577

[76] Du X, Wang Y, Zhang X, Ma G, Liu Y, Wang B, et al. A study of plant growth regulators detection based on terahertz time-domain spectroscopy and density functional theory. RSC Advances. 2021;**11**(46):28898-28907